Ethanol Fuel

Learn to make and use ethanol to power your vehicles

Samuel Edison & Alan Delfin

medical or professional advice. The content of this book has been derived from various sources. Please consult a licensed professional before attempting any techniques outlined in this book.

By reading this document, the reader agrees that under no circumstances is the author responsible for any losses, direct or indirect, which are incurred as a result of the use of information contained within this document, including, but not limited to, —errors, omissions, or inaccuracies.

Somebody, please press the world's
reset button

TABLE OF CONTENT

ABSTRACT

There is a lot of different fuel you can make use, but the fact that you are reading this reveals your interest in ethanol as a fuel. In the world today, petroleum, solar, biodiesel are the common fuel people use today. Others may be considering ethanol because of its benefit to replace other fuel in cars, generators, tractor, or to power farm equipment or home.

The majority of people who read this book to the end will likely have learned a lot about ethanol, such include; history of ethanol, how to produce ethanol at home, uses of ethanol, application of ethanol and so on. Ethanol has a long history, certainly as a beverage but also as a fuel which only comes to use in the 19th century for lighting.

The path of ethanol from a light source to fuel, an additive for pure air testing and bridging technology enable us to move into an oil-free era which is exciting. The story is full of political issues, the effects of wars, industrial espionage and the pure energy of a popular movement. The most important story, however, is the fact that a full litany of common carbohydrates, not just food crops, but also agricultural slaughter, food waste, and plants that are normally bothersome, can actually become a viable fuel that is effectively distributed or produced.

According to the approach adopted for the production of ethanol, it is entirely possible to maintain a fully autonomous, self-sustaining and environmentally responsible operation that produces not only fuel but also valuable by-products that can be sold, replaced, or recycled. In this way, ethanol has real advantages over other renewable fuels because it does not need many processes. Releasing unpredictable changes in traditional commodity markets can be a real advantage in long-term planning, insurance, and peace of mind.

1 INTRODUCTION TO ETHANOL

1.0 WHAT IS ALCOHOL AS A FUEL?

Alcohols are used for fuel. Taking consideration of this first aliphatic alcohol (methanol, ethanol, propanol, and butanol) is of interest as far as fuel production. Methanol, ethanol, propanol and butanol can be synthesized chemically or biologically, it also have characteristic which enable them to be used in the internal combustion engines. The general chemical formula for alcohol is $C_nH_{2n+1}OH$.

Mostly, methanol is produced from natural gas, although some methanol is produced from biomass using a similar chemical process. Ethanol is commonly produced from biological material using the fermentation process. Their high octane ratings gave them the benefit of increasing their fuel efficiency which largely offsets the lower energy density compared to gasoline or diesel fuel, thus results to fuel economy in terms of a kilometer per liter.

Using alcohol as fuel is not new. When Nikolaus Otto invented the internal combustion engine in 1872, there was no gasoline available; the indicated fuel was ethyl alcohol at 180-190. Ford's "T" model was developed for both use gasoline and alcohol.

Alcohols in general and ethanol in particular, are excellent fuels for cars, machinery. The reason that the alcohol-based fuel was not fully utilized is that gasoline was once cheap, available and easy to produce.

However, crude oil is getting scarce and the historical price difference between alcohol and petrol is decreasing. At the moment considerable efforts are being made to find and develop alternative energy sources to obtain reserves of oil and other decreasing fossil fuels. Edward Teller, one of the greatest physicists in the country, points out: "There is no single recipe for solving the energy problem. Energy conservation is not enough, petroleum is not enough, Coal is not enough, Nuclear energy is not enough, geothermal energy is not enough and developments alone will not be enough, only the right combination of these factors is sufficient. "

Alcohol can be an important part of the solution, but certainly not a panacea. If all the existing agricultural surpluses are converted to ethanol, alcohol accounts for less than 5% of our fuel need. The ability to convert cellulose residues to ethanol and general biomass into methanol and the most optimistic total remains below 10% of our current needs! However, this is a very important rate of 5 or 10%, since it can be renewed every year and saves a gallon of fat from each gallon of oil produced.

1.1   Chemical composition

Alcohol and gasoline, despite the fact that they are from different chemical classes, are remarkably similar. Gasoline is mostly a mixture of "hydrocarbons". Hydrocarbons are a group of chemical substances composed exclusively of carbon and hydrogen atoms. This is a very large chemical class containing many thousands of substances.

Most of the fuels we use such as coal, gasoline, kerosene, fuel oil, butane, propane, etc. are chiefly hydrocarbons. Referring to Figure below, the simplest member of this group is methane which consists of a single carbon atom and four hydrogen atoms. Next is ethane with two carbons and six hydrogen. Propane has three carbons and butane has four. The substances just named are gases under ordinary conditions. As we add more carbons to the hydrocarbon molecule, the chemicals formed become liquids: pentane, hexane, heptane, octane and so on. As we continue with even more complex molecules, the substances get progressively oilier, waxier and finally solid.

SIMPLE (ALIPHATIC) HYDROCARBONS:

```
        H               H  H              H  H  H
        |               |  |              |  |  |
    H-C-H           H-C-C-H           H-C-C-C-H
        |               |  |              |  |  |
        H               H  H              H  H  H
    METHANE           ETHANE             PROPANE


    H  H  H  H          H  H  H  H  H  H  H  H
    |  |  |  |          |  |  |  |  |  |  |  |
H-C-C-C-C-H      H-C-C-C-C-C-C-C-C-H
    |  |  |  |          |  |  |  |  |  |  |  |
    H  H  H  H          H  H  H  H  H  H  H  H
      BUTANE                   OCTANE
```

SIMPLE (ALIPHATIC) ALCOHOLS:

```
        H               H  H              H  H  H
        |               |  |              |  |  |
    H-C-OH          H-C-C-OH          H-C-C-C-OH
        |               |  |              |  |  |
        H               H  H              H  H  H
    METHANOL          ETHANOL           n-PROPANOL


    H  H  H  H  H  H  H  H  H  H  H  H  H  H  H  H
    |  |  |  |  |  |  |  |  |  |  |  |  |  |  |  |
H-C-C-C-C-C-C-C-C-C-C-C-C-C-C-C-C-OH
    |  |  |  |  |  |  |  |  |  |  |  |  |  |  |  |
    H  H  H  H  H  H  H  H  H  H  H  H  H  H  H  H
                      CETYL ALCOHOL
```

Chemical structure

1.3 Combustion properties

One of the most important properties of a fuel is the amount of energy obtained from it when it is burned. Referring to Figure below, note that the hydrocarbon octane, which represents an "ideal" gasoline, contains no oxygen. In comparison, all of the alcohols contain an oxygen atom bonded to a hydrogen atom in the hydroxyl radical. When the alcohol is burned, the hydroxyl combines with a hydrogen atom to form a molecule of water. Thus, the oxygen contained in the alcohol contributes nothing to the fuel value.

	TYPICAL REGULAR GASOLINE	OCTANE*	METHYL ALCOHOL	ETHYL ALCOHOL
Chemical Formula	Complex	C_8H_{18}	CH_3OH	C_2H_5OH
Molecular Weight	Complex	114	32	46
Heating Value (Btu/lb)				
High Value	20,250	20,570	9,770	12,780
Low Value	19,000	19,080	8,640	11,550
Latent Heat of Vaporization (Btu/lb)	140	141	474	361
Specific Gravity (@60°F)	0.745	0.702	0.796	0.794
Stoichiometric Ratio	15:1	15.1:1	6.45:1	9:1
Boiling Temperature (°F)	100-400	258.2	148.5	173.3
Octane Number (Research)	80	100	106	106
Energy of Stoichiometric Mixture (Btu/ft^3)	94.8	95.4	94.5	94.7

*Can be considered as "ideal" high-test gasoline

Physical properties of alcohol and gasoline

The relative atomic weights of the atoms

involved are: hydrogen, 1; carbon, 12; and oxygen, 16. Since methyl alcohol has an atomic weight of 32, half the molecule cannot be "burned" and does not contribute any fuel value. As expected, methanol has less than half the heat value (expressed in Btu/lb) of gasoline. Ethanol, with 35% oxygen, is slightly better with 60% of the heat value of gasoline. If the heating value of methyl and ethyl alcohol were considered alone, they would appear to be poor choices as motor fuels.

However, other redeeming qualities such as "latent heat of vaporization" and anti-knock values make alcohol fuels superior, in some ways, to gasoline. When a fuel is burned, a certain amount of air is required for complete combustion. When the quantity of air and the quantity of fuel are exactly balanced, the fuel air mixture is said to be "stoichiometrically" correct. Again referring to Figure above, the stoichiometric ratio for gasoline is 15:1 or 15 pounds of air for each pound of gasoline. The figures for methyl and ethyl alcohol are 6.45:1 and 9:1 respectively. On a practical level, this means that to burn alcohol effectively, the fuel jets in the carburetor must be changed or adjusted to provide 2.3 pounds of methanol or 1.66 pounds of ethanol for each 15 pounds of air. Referring to the last entry in Figure 2-2, an interesting fact is that if we provide the correct stoichiometric mixture and then compare on the basis of the energy (in Btu's) contained in each cubic foot of the different fuel/air mixtures, the fuels are almost identical: gasoline 94.8 Btu per 8 cubic foot; methanol 94.5 and ethanol 94.7! This means that gasoline and alcohol are about equal in what is called "volumetric efficiency" when burned in a correctly adjusted engine

1.4   OCTANE RATINGS

If a certain fuel is burned in an engine in which the compression ratio can be varied and this ratio is gradually increased, a point will be reached when the fuel will detonate prematurely. This is because as a gas is compressed, heat is generated. If the explosive fuel/air mixture in an engine cylinder is compressed enough, the resulting heat will cause it to detonate. Since gasoline engines are designed so that the mixture is detonated by the spark plug at the beginning of the downward movement of the piston following the compression stroke, preignition or "knock" occurring during the compression stroke is undesirable.

Indeed, severe knock can quickly overstress and destroy an engine. Since greater compressionratios in an engine mean increased power per stroke and greater efficiency, the ability of a fuel to resist premature detonation is a desirable quality. The "octane" numbers assigned to fuels are based on the pure hydrocarbon, octane, which is considered to be 100. At the other end of the scale, n-heptane is considered to have an octane rating of zero. The octane number of an unknown fuel is based on the percentage volume of a mixture of octane and nheptane that matches it in preignition characteristics.

In practice, these tests are conducted in a special test engine with variable compression. Alcohols have a relatively high anti-knock or octane rating. As noted in Figure 2-3, alcohols have the ability to raise considerably the octane ratings of gasoline with which they are mixed. The effect is greatest on the poorer grades of gasoline. A 25% blend of ethanol and 40 octane gasoline will have a net increase of almost 30 points! This increase is one of the major advantages of "gasohol".

The ability to increase octane rating means that: (1) a lower (therefore cheaper) grade of gasolinecan be used to obtain a fuel with a certain octane rating; and (2) the use of traditional pollution producing antiknock additives such as tetraethyl lead can be eliminated. The addition of about 10-15% ethanol to unleaded gasoline raises the octane rating enough so that it can be burned in high compression engines that previously could not use unleaded fuel. This use of ethanol is not new, of course, because ethanol was the original gasoline additive for increasing the octane rating. The term "ethyl" used to describe a high-test gasoline comes from ethyl alcohol, not tetraethyl lead!

Journeytoforever.org.(2019).   *[Online]*   *retrieve*   *from:* *http://journeytoforever.org/bflpics/AlcFuelManual.pdf*

1.5   Alcohol as a fuel

The idea of using alcohol as a fuel in cars is not new. At the launch of the T model, Henry Ford predicted he would work with alcohol produced from renewable sources. Brazil uses cane liquor for many years. Because of this experience, alcohol is considered an alternative to fossil fuels. The book is a compilation on the production of methanol and ethanol, the use of alcohol blends in automotiveapplications and the use of alcohols in a fuel cell. The chapters were created by about 26 American and Europe authors individually, reflecting the broad scope of the book and the international dimension of the subject. The future availability of oil, its security of supply and, increasingly, its impact on climate change have motivated the search for alternative sources of energy, especially sustainable, low impact, or even without environmental impacts.

Alcohol from renewable sources is one of the main competitors to meet these challenges. To foster these changes, government initiatives and tax incentives in the US, Brazil, Europe, and the Far East not only stimulate the development and production of alcohol, but also the gradual introduction of alcohol fuels mixed with fossils. This book is therefore ideally suited to give for the first time a comprehensive overview of the production of alcohols, their use as fuel in internal combustion engines as well as more advanced concepts such as fuel cells and portable mobile applications through fuel cells. The book will appeal to a broad readership. People who are not yet familiar with this topic, but who are interested in the development of renewable fuels, will find great value in this book, even if they do not follow the more technical chapters on biochemistry and catalyst development for fuel cell applications. Likewise, persons who are familiar with or interested in certain areas, e.g. For example, the challenge of developing catalysts for direct alcohol fuel cells, the book and its many references of particular value.

1.6   Why is Ethanol Important?

U.S. Ethanol Industry Has Grown by Leaps and Bounds

The U.S. ethanol industry has grown from a handful of small plants producing 175 million gallons in 1980 to 210 plants in 27 states producing a record 16.1 billion gallons of ethanol in 2018. Today, we make up slightly more than 10 percent of the U.S. gasoline supply.

Ethanol Plants are Creating Jobs and Helping Fuel the U.S. Economy

In 2018, the industry directly employed 71,367 American workers and supported an additional 294,516 indirect and induced jobs across the economy. Nearly one in four of those workers is a veteran of the U.S. military. The ethanol industry generated $46 billion in gross domestic product last year and boosted household income by $25 billion.

Boosting Rural Economies

Ethanol and feed co-product production provide a valuable market for corn grown in the United States. A typical dry mill ethanol plant adds nearly $2 of additional value – or 55 percent – to every bushel of corn processed.

Driving Energy Independence

Ethanol is a win for all Americans. It's more affordable than traditional gasoline, reduces harmful vehicle emissions, supports nearly 300,000 American jobs and protects America's energy independence. In 2018, ethanol displaced nearly 600 million barrels of crude oil and petroleum import dependence fell to just 14 percent compared to 60 percent in 2005. Without ethanol, petroleum import dependence would have been 20 percent in 2018.

Producing Food AND Fuel

Ethanol bio refineries make more than fuel; they also generate highly nutritious animal feed like distillers grains. One-third of every bushel processed by a plant is used to make animal feed, resulting in the production of more than 41.3 million metric tons of feed in 2018 alone. The low cost and nutritional properties of distiller's grains make it one of the most sought-after feed ingredients in the world.

A Cleaner, Greener Fuel

Ethanol is responsible for removing the carbon equivalent of 20 million cars from the road. At the same time, the environmental impacts of producing ethanol have been greatly reduced. Natural gas and electricity use at dry mill ethanol plants has fallen nearly 40 percent since 1995, while consumptive water use has been cut in half. This has occurred while the amount of ethanol produced from a bushel has increased. Producers are getting 15 percent more ethanol from a bushel of corn than 20 years ago. The resulta smaller carbon footprint and increase in energy efficiency. Ethanol use reduces greenhouse gas emissions by 40-45 percent compared to gasoline– even when hypothetical land use change emissions are included. By displacing hydrocarbon substances like aromatics in gasoline, ethanol also helps reduce emissions of air toxics, particulate matter, carbon monoxide, nitrous oxides, and exhaust hydrocarbons.

Boosting Efficiency

Today's producers get more ethanol out of every bushel. On average, 1 bushel of corn (56 pounds) processed by a dry mill ethanol plant produces: 2.86 gallons of denatured fuel ethanol; 15.9 pounds of distillers grains animal feed; 0.75 pounds of corn distillers oil; and 16.5 pounds of biogenic carbon dioxide Providing More Consumer Options We know Americans are looking for more competition and greater savings for the fuel that powers their vehicles. Nearly all U.S. gasoline today contains 10 percent ethanol and the use of 15 percent ethanol blends and flexes fuels like E85 is increasing. 15 percent ethanol blends (E15) are higher quality fuels that offer greater savings. Today, E15 is available at over 1,500 stations in more than half the country. The EPA has approved E15 use in more than 90 percent of the existing U.S. auto fleet, and 9 out of 10 new cars carry the manufacturer's warranty and approval for E15. Beyond E10 and E15, flex fuels like E85 (85 percent ethanol) are sold at more than 4,500 retail stations.

Expanding Global Markets

The United States is the world's leading ethanol producer. Even in the face of new trade barriers in key markets, U.S. ethanol exports surged to a new record of more than 1.6 billion gallons in 2018. U.S. ethanol was shipped to Canada, Brazil, and India, South Korea and the Philippines and many other countries in 2018.

Future Innovation

Today's ethanol biorefinery operates much like a chemical refinery, able to produce multiple renewable fuels and products. Some bio refineries are producing biodiesel and renewable diesel from corndistillers' oil, but the largest impact has been in corn kernel fiber production. The addition of "bolt-on" technologies has allowed ethanol producers to expand yields by processing ethanol from corn fiber, a cellulosic portion of the grain. Unleashing corn kernel fiber ethanol production could result in existing ethanol plants producing hundreds of millions of gallons of cellulosic ethanol

Why is Ethanol Important? – Renewable Fuels Association. (n.d.). Retrieved from: https://ethanolrfa.org/consumers/why-is-ethanol

1.7   Gasohol and E85

The Oil Crisis, which began in 1973-74 and was triggered by the decision of the Organization of the Arab Petroleum Exporting Countries (OAPEC) to stop delivering oil to countries that supported Israel in October 1973 in the Yom Kippur War, was directly from the US Economy affected. The embargo was lifted in March 1974, but the approval of the Organization of Petroleum Exporting Countries (OPEC) to raise world oil prices quadrupled on quadruple within a few months. Several legislative measures immediately investigated the use of cellulose and biomass in the production of alcohol and corn-producing countries that prefer ethanol and gasoline to reduce the consumption of imported oil.

The original attempts to market ethanol and gasoline were mixed with wholesalers and consumers, but the 1978 Energy Tax Act supported the growing movement by recognizing and defining gasoline as a "blend of at least 10% petrol by volume", excluding alcohol derived from oil, natural gas or coal ", which provides for renewable raw materials and supports the use of alcohol as a fuel in this part of the state excise duty on petrol.

When another oil crisis occurred in 1979, Chevron, Amoco, Texaco and other commercial fuels were sold. Gasohol retail has been promoted in some markets by marking individual service stations and the E-10 label. At the same time that the United States began leaking lead in gasoline (in 1986 a total ban was imposed), ethanol became more attractive than the octane count.

At that time, there were only a handful of commercial ethanol producers that produced approximately 50 million gallons of fuel alcohol per year. After several federal subsidies and a tariff on imported ethanol, domestic ethanol production increased to almost 600 million gallons in the mid-1980s. Although the price of crude oil fell, with less competitive ethanol, environmental concerns about carbon monoxide concentrations focused on the use of oxidized fuels.

Ethanol was a viable oxidation, as were methyl tertiary l butyl ether (MTBE) and ethyl tertiary butyl ethyl ether (ETBE). Of these, only ETBE used ethanol in manufacturing; MTBE is made from natural gas and oil and which has dominated the market because of its lower costs. In 1992, the Clean Air Act was to use oxidized fuels in 39 of the most significant "non-attainment" of carbon monoxide, followed three years later in nine regions characterized by strong noncompliance with the ozone layer. Gasoline has been redesigned in many urban areas of the country, including products containing oxygen, mainly MTBE.

However, in the late 1990s, traces of MTBE began to appear in drinking water sources and some countries began to ban the use of MTBE in fuels by changing the demand for ethanol and ETBE. In 2000, the Environmental Protection Agency recommended the national phase-out of MTBE, but countries that were important users of the oxidized product nevertheless advanced in this direction. The main refineries quickly decided to voluntarily use ethanol-based oxidized compounds in their reformulated gasoline, fearing possible legal action because of the carcinogenic effects of MTBE.

The production of E-85 as a separate gasoline blend began in the early 1990's with an energy policy that included an alternative fuel containing 85% ethanol and 15% hydrogen of gasoline. In 1997, major automakers began volume production of "Flex Fuel" vehicles developed for a combination of unleaded gasoline and E-85 or both. So far, there are 6.8 million Flex Fuel vehicles, including cars, and vans.

Flex Fuel vehicles are low priced compared to petrol gasoline vehicles and are factory-guaranteed and approved by the EPA. The main difference between the Flex Fuel vehicle and the conventional gasoline vehicle is the addition of a fuel sensor that detects ethanol and gasoline. Other components such as the fuel tank, the fuel hoses and the fuel injection system have been slightly modified. The Electronic Vehicle Controller / Transmission Control Module (ECM / PCM) Card is being reprogrammed to expand the fuel spectrum. The fuel consumption of Flex Fuel vehicles can be reduced by using the E-85, as the engines are not just optimized for alcohol.

In general, the price of E-85 is lower than the price of unleaded prices, which compensates the difference. The product made from agricultural products is considered a renewable fuel and has excellent impact resistance thanks to its octane rating of 105. From an environmental point of view, E85 reduces greenhouse gas emissions by around 20% compared to unleaded gasoline and also reduce evaporation and carbon monoxide emissions, as well as toxic emissions such as benzene and other carcinogens known. The E-85 contains 15% unleaded gasoline to allow Flex Fuel vehicles to start in cold weather.

In particularly cold climates, the Winter E-85 may contain a little more fuel to allow a cold start. Even though the E-85 program is still called E-85, some ethanol representatives argue that the entire program is simply a concession to automakers and the oil industry to install a cold start system beyond the modest additional costs of an installation. They say it would have been as convenient to produce E-100 (Pure Ethanol) and Flex Fuel vehicles that can be used from all types of pure ethanol and gasoline, such as Brazil.

R. F. (2009). *Alcohol Fuel A Guide to Making and Using Ethanol as a Renewable Fuel*. Canada, New York: New Society. Retrieve from: www.newsociety.com

1.8   Food vs. Fuel

None of the anxiety over rising corn prices escaped the attention of the nation's food industry, whose leaders expressed concerns over diverting grains to fuel use and the subsequent effect it might have on food staples and livestock feed. True to form, costs of grain-based products rose noticeably, a boon for price starved farmers, but a bust for consumers of everything from mea t to soda pop to dog food — anything that uses corn sweeteners in its manufacture.

Moreover, the demand for corn has spurred a large agricultural interest in production. (About 20 percent of the corn crop goes into the manufacture of ethanol.) Since corn is an especially nitrogen-fertilizer intensive crop, the inevitable runoff from farm fields has been linked to algae blooms in the nation's waterways, some widespread and intensive to the point of affecting marine life and subsequent seafood harvests.

On a larger scale, the demand for corn and other biofuel crops has prompted further agricultural development globally, with land clearing, deforestation and unprecedented water usage the result.

Despite President Bush's State-of-the-Union appeal, the most expedient way to make ethanol at this point in time is with corn, not with wood chips or switch grass as he envisioned. But there's a caveat: All of this involves massive market-scale production—manufacture on a level so enormous as to apply only to entities entrenched in the world of Agribusiness. At this point, making ethanol is so profitable —thanks to government subsidies and unusually high oil prices—that ethanol plants can't be built quickly enough to meet demand. And that will continue as long as fuel prices are high and corn prices remain relatively low.

Yet what many people are talking about is far removed from this. It is fuel production on an appropriate local level, either cooperatively or within a farm system that has access to a variety of reliable feedstock. At some point, the price of petroleum, or the cost of refining it or delivering it, may be so high as to make small scale production of farm ethanol very attractive to those who seek to be in control of their own fortunes. Going one step further, acquiring the raw materials for fuel from a dependable local source — be it a product of the area economy, a crop chosen for its conversion value, or agricultural waste — offers a degree of separation from an uncertain and perhaps unstable national economy, which holds a certain amount of appeal and even comfort.

There was a time, well within the memories of some folks living today, when the energy used to run a farm or homestead came from the farm itself. The food was grown there, the wood that cooked the food and heated the house was felled there, and the muscle that planted and harvested the crops was fed on the hay and grasses raised there. Wind pumped the water, and often provided basic electrical needs. Today's technology can only improve on that.

Things may yet come full circle for those who plan. The ethanol made from farm crops can power the tractors, fuel the furnaces, run the generators and feed the livestock with protein-rich distiller's grains. By-products from ethanol production can fire the cookers that boil the mash and heat the distillation columns. And the labor can remain on the farm and in the community it supports, rather than left to the uncertainties of an unsure outside market.

Journeytoforever.org.   (2019).   [Online]   retrieve   from:
http://journeytoforever.org/bflpics/AlcFuelManual.pdf

2 ETHANOL   REGULATORY   GUIDE

2.0   How to Apply For a Fuel Alcohol Permit

In order to legally make fuel alcohol one must have the proper permits. A federal fuel alcohol producer permit is available for free from the TTB. The application must be submitted and approved by the TTB before fuel alcohol is manufactured. The TTB is the division of the government that is in charge of regulating and overseeing distillation. TTB stands for Alcohol and Tobacco Tax and Trade Bureau which is part of the U.S. Department of the Treasury.

Even though the Federal Government allows the manufacture of fuel alcohol with a permit it is important to realize that some states don't allow fuel alcohol production. State distilling laws vary from state to state. Some states have no laws on owning a still, but prohibit the distillation of alcohol while other states prohibit possession of a still unless it's for fuel alcohol. States such as North Carolina requires a state fuel alcohol permit as well as the federal fuel permit. Some states may prohibit possession of distillation equipment and distilling altogether. It is important to research local state laws before applying for a federal fuel permit.

Download the federal fuelpermit here(http://www.ttb.gov/forms/f511074.pdf) and follow along with our guide to properly fill out the alcohol fuel plant application. Our guide will focus on the small fuel alcohol production plant as production with a Clawhammer Supply; LLC still will produce less than 10,000 proof gallons a year.

1) Type of Plant: If using a Clawhammer Supply still check the Small – 10,000 Proof Gallons or fewer boxes.

2) Amended Permit: Leave this blank when applying for a new permit

3) Name of Owner: Type in first / middle/ and last name

4) Daytime Telephone Number: Enter full phone number

5) EIN or SSN- Enter your Social Security Number or if registered as businesses use the EIN- Employer Identification Number

6) DOB: Enter the DOB for the applicants or applicants.

7) Location: Type in the street name where the small distillation plant will be located.

8) Mailing address: If the mailing address is different than the plant locations enter it here. If the mailing address is the same as the plant locations leave it blank.

9) Premise for Alcohol Fuel Plant Are: If you own the premise check owned by the applicant and skip to number 11.

10) If the premise is rented fill out section 10 which must be signed by the property owner.

11) Stills For Fuel Production On Plant Premises:

a) Still Manufacturer: Clawhammer Supply, LLC

b) Serial Number: The Serial Number for Clawhammer Supply, LLC is the order number

c) Kind of Still: Pot Still

d) Capacity – The Clawhammer Supply, LLC

still is a pot still and a pot still may be shown by giving the wine gallon capacity of the pot still.

a. The 1 gallon Clawhammer Supply, LLC still 1 gallon

b. The 5 gallon Clawhammer Supply, LLC still 5 gallons

c. The 10 gallon Clawhammer Supply, LLC still 10 gallons

12) Basic Materials to be used in production of spirits: If you are planning on using a variety of fermentables select the appropriate boxes. I selected grain, sugar based crops, and fruit or fruit products as these are the fermentables I use when making a mash.

13) Security Measures: On our property we have a fenced in yard and a distilled spirits building which is locked.

14) Diagram of plant premise: The diagram of your plant premises may be drawn by hand and does not have to be drawn to scale.

15) I will comply with the clean water act: Check I will comply with the clean water act and don't discharge into navigable water s of the U.S. Navigable waters are waterways that provide a

channel for commerce and transportation of people and goods.

16) Skip: Only applies to Medium and Large Alcohol fuel plant applications

17) Sign the form

18) Title: Owner

19) Enter the date

Print out 3 copies of the TTB application and any attachments (the drawing of the premise); make sure the attachments are labeled with first and last name and are printed on the same size paper as the application.

Mail the completed application to: Director, National Revenue Center, 550 Main St Ste 8002, Cincinnati, OH 45202-5215. If required by your state, submit a copy of your approved application to the alcohol beverage agency or other State agency. It took 3 months to receive our fuel alcohol permit.

Fuel Alcohol Permit: The Ultimate Guide. (n.d.). Retrieved from https://www.clawhammersupply.com/blogs/moonshine-still-blog/25856388-fuel

2.1   Hazard and Safety

Hazard

1. Over pressurization; explosion of boiler
2. Scalding from steam gasket leaks
3. Contact burns from steam lines
4. Ignition of ethanol leaks/fumes or grain dust
5. Handling acid bases
6. Suffocation

Safety precaution

- Regularly maintained/checked safety boiler "pop" valves set to relieve when pressure exceeds the maximum safe pressure of the boiler or delivery lines.
- Strict adherence to boiler manufacturer's operating procedure.
- If boiler pressure exceeds 20 psi, acquire ASME boiler operator certification. Continuous operator attendance required during boiler operation.
- Place baffles around flanges to direct steam jets away from operating areas. (Option) Use welded joints in all steam delivery lines.

- Insulate all steam delivery lines.
- If electric pump motors are used, use fully enclosed explosion-proof motors.
- (Option) Use hydraulic pump drives; main hydraulic pump and reservoir should be physically isolated from ethanol tanks, dehydration section, distillation columns, and condenser.
- Fully ground all equipment to prevent static electricity build-up.
- Never smoke or strike matches around ethanol tanks, dehydration section, distillation columns, and condenser.
- Never use metal grinders, cutting torches, welders, etc. around systems or equipment containing ethanol. Flush and vent all vessels prior to performing any of these operations.
- Never breathe the fumes of concentrated acids or bases.
- Never store concentrated acids in carbon steel containers.
- Mix or dilute acids and bases slowly-allow heat of mixing to dissipate.

- Immediately flush skin exposed to acid or base with copious quantities of water.
- Wear goggles whenever handling concentrated acids or bases; flush eyes with water and immediately call physician if any gets in eyes.
- Do not store acids or bases overhead work areas or equipment.
- Do not carry acids or bases in open buckets.
- Select proper materials of construction for all acid or base storage containers, delivery aides, valves, etc.
- Never enter the fermenters, beer well, or stillage tank unless they are properly vented

2.2   fire Code

Compliance with local fire regulations is legal and essential for all production facilities. Unfortunately, it can also be a tool to exert forces if this is not the case. The presence of commercial or micro-industrial activities can cause problems even in suburban areas. Manufacturing industrial neighbors does not normally enhance the value of residential real estate. At the same time, however, this is usually based on a scale.

It is unlikely that very little activity in the backyard garage will cause a problem unless someone complains about noise or odor, which should not matter if you plan your facility properly. Before working in real time or investing in serious activities, it is advisable to study the land use and development laws in the area in which you operate. In rural areas, the restrictions should be almost non-existent and the space is not significantly different from a typical farm.

In urban areas, even a small distillery may be necessary, and certain types of production may be prohibited. It must be prepared for electrical and water installations and fire tests. Speaking of fire, with every standard measure, it is safer to ignite ethanol than gasoline. Gasoline evaporates and ignites at a much lower temperature. But there are two things that could arouse the concern of your Commissioner of the fire department which are:

1. Burns with an almost invisible flame, even with less evidence (this characteristic changes when the fuel is denatured)

2. The foams used in traditional gasoline blends are dissolved in alcohol.

However, plain water is never used in oil burns because it spreads the flames; it works against alcohol that has enough volume to deepen the fuel deeply. Conventional fire extinguishers of type A B C are also used to remove small pieces of alcohol. In addition, special polymer foams are available to extinguish the main sources of fire. The reluctance of fire safety professionals to support alcohol fuel is due to the production of larger industrial production plants, where tank trucks and rail vehicles are filled with dozens of liters of alcohol. With safe storage techniques (above-ground tanks, adequate ventilation and sealed seals) there are several hundred liters of high-risk fuel alcohol.

2.3 Insurance

The AFP federal license is not insurable, but is recommended for those who rent or pay a mortgage or for those who use the non-owners for production or commercial package. It should be remembered that the production of alcohol and fuel is not fundamentally dangerous just because it is a fuel, especially compared to gasoline. The auto-ignition temperature for pure ethanol (100%) is 793 ° F compared to about 430 ° F for gasoline. Its flashpoint is also 100 degrees higher and the latent heat of vaporization of ethanol, which measures the amount of energy needed to vaporize it, has more than doubled. The fuel factor and boiler lighting method, however, may affect traditional practices.

2.4   Batches record keeping

The registration and measurement of alcohol content are described in the Federal Code (CFR) mentioned above. If necessary, the rules vary between beverage alcohol and ethanol fuel (when denatured) can be given in gallons of wine. Therefore, it is not technically necessary to determine the detection of the ethanol produced.

However, you must make accurate measurements of the strength and volume of alcohol you produce. Therefore, you will probably need a precise thermometer, hydrometer and laboratory cylinder to perform periodic corrective actions, especially during the storage of an un-denatured product. (The ethanol temperature affects the readings, so the table in Appendix A can be used as actual proof).

The licensee of the alcohol manufacturer must submit activity reports, but small farmers must report annually only within 30 days of December 31 of each calendar year. The alcohol fuel Report F 5110.75 requests information on the test facilities for the liquors produced the quantity and type of material used as denaturing agent and the production of the wine produced in the wine-growing zone.

2.5 Denaturalizing alcohol to be used as fuel

Before distilled spirits can be disposed of in a factory, they must not be suitable for use by the addition of unleaded petrol, petroleum, methyl isobutyl ketone or any other Bureau approved substance. The TTB has approved at least eight denaturing formulas (listed in Part 27 of the Federal Code, Part 21) and 27 CFR 21.91 provides substitute materials upon request, subject to the Director's written consent.

In April 2008, the TTB authorized the use of "direct competition" as a denaturant for alcohol fuel manufacturers due to their lower costs. The gasoline or direct-launched mixture is defined by the APC and consists mainly of cracking or gasoline condensate.

The denaturing formulas are based on 100 gallons of alcohol units.

For most small scale producers of ethanol fuel, the use of unleaded petrol or kerosene is an optional denaturant because it is the simplest. It is necessary to add at least 2 liters of denaturing agent per 100 liters of petrol containing at least 195% alcohol. In the second stage, the Bureau published a brochure on "Fuel consumption distillation fuel" from ATFP-5000.5, summarizing these data and other information on the preservation of registrations, bonds, etc. This information is now available online at *ecfr.gpoaccess.gov* so you can search for terms by chapters, periodically, and partially.

R. F. (2009). *Alcohol Fuel A Guide to Making and Using Ethanol as a Renewable Fuel.* Canada, New York: New Society. Retrieve from: www.newsociety.com

3   Basics   of   production

3.0   Small scale production

Ethanol is like any renewable product, is a commodity whose value is determined by the content. As a small scale ethanol fuel producer, your motivation can be more than economical-you can reduce your carbon footprint by using renewable source or becoming more self-sufficient. However, you are still subjected to the same process that manipulates the fuel market for everything else you buy.

Well this doesn't mean you can't control your own price of ethanol fuel product. Is just that you will have to accept the realities and make some changes in the energy balance to favor you. Ethanol fuel production can be an inspiring and even stimulating experience but also it takes a lot of work for the production. The obligation to determine the level of a private distillery is not an informal enterprise and the time required to purchase the raw materials and operate the equipment is significantly considerable.

The economies of scale tend to bedevil everyone in planning a small operation because so many variables are entwined. How do you calculate how much fuel you'll need? Should you convert one vehicle for straight ethanol use or try an E-85 blend? If you'd rather start small, how efficient can a smaller operation be? Would it be better to go with a larger still to keep costs down? Will your feedstock and fuel sources be consistent? If not, how will that change ethanol output and costs?

None of these questions have a definitive answer because each is dependent on some other part of the operation. Fortunately, as a small player, you'll enjoy a flexibility that larger operations don't have, and you'll also be somewhat freer to take more of a "seat of the pants" approach to your effort.

Ideally, your ethanol plant would be part of a farm or market-growing venture, for two reasons. First, as a grower you'd already have a familiarity with the day-to-day practices that agriculture entails. This includes working within a routine, searching for markets, dealing with equipment in both fair and inclement weather, and quite importantly, improvising when necessary to keep things running smoothly. As anyone who has worked the land can tell you, the most successful farmers are well-rounded Renaissance people who can roll with the punches and take things in stride.

Second, a working farm provides a readymade outlet for the manufactured fuel and its by-products. Most any internal-combustion engine or heating appliances can be adapted to run on alcohol — this inventory includes tractors, trucks, pumps, generators, burners and furnaces—and the residual material from mash production contains enough nutrient to supplement normal livestock feed.

If agriculture is not in your background, it's still possible to manufacture alcohol, even economically, provided you have are liable source of raw material, or feedstock. As you'll see in the next chapter, there are many viable candidates for ethanol production, including both sugar and starch crops. Residues from canning and juicing operations, even far from the farm, are also distinct possibilities. Realistically, it would be difficult to carry on much more than an experimental venture in a confined space such a suburban backyard, but it's still possible. Ideally, a rural setting or a location where there's room to expand and function without interference would be the better choice.

3.1 Sourcing raw materials

Finding a reliable and consistent power source can be a challenge. In the next chapter i will discuss the distinction between sugar crops such as sugar cane, sugar beet juice and starch crops such as maize, sorghum, cereals and potatoes. Suffice it to say that some plants produce more starch or sugar per ton or per acre than others and that at a reasonable cost, crops with more concentrated nutrients is the best option.

More complicated, however, is the fact that the equipment needed to process the raw material varies depending on the crop. The flour mill is very different from the extraction equipment used to treat sugar beet. If you cannot develop a responsible resource, it would not be wise to invest in certain equipment. Consider renting (or renting) this equipment or using the services of a local co-op.

If you live in a rural community with breeding and packaging house and companies, you may find that recovery and deterioration are the best economic option. If properly handled, most cooperatives and private refiners should be prepared to negotiate an attractive contract, a contract that allows them to measure the value of their change as a raw material for performance results over a given period of time. In the same way, you can always try to negotiate agreements with individual farmers, possibly in exchange for the disposal of field and fruit waste, which at least gives you the raw material you need temporarily during the determination of profitability.

Storage may be a problem in some cultures. Some products must be handled within a few months of harvesting; otherwise they must be dried sufficiently to preserve them. Drying and storage cause additional costs and should be avoided at the same time. Of course, you must take precautions to keep your cargo daily, especially if you plan to use the batch continuously instead of using it continuously.

3.2   Buy or build it equipment

For a small scale ethanol fuel manufacturer, many models are so justified that designing equipment is much easier and cheaper than buying them. This is especially true for low capacity. No expensive stainless steel components are required on this scale: a standard mild steel tube makes the columns and water pipes, and in some applications a plastic pipe can be used. In addition, tanks and vats do not have to be special, but it is often cheaper to buy equipment used in an agricultural auction (storage of dairies and stainless steel processes or tanks are common auctions).

If you have welding skills and a place to carry out your operation, you are ahead of the game. For parties that do not exist, there is no reason to use new materials. Any metal yard will likely produce the parts you need. If you are not ashamed or fussy, the old oil tank can make a real boiler vat, and the corresponding liquidtanks can be used as a mixed steam generator. Many components are tasks of other applications, so you should have a creative eye when buying. Unfortunately, many of the fabricated steel products, especially stainless steel, have added value in their second-hand markets, as foreign markets for quality steelmaking have generally increased, especially well-produced US products, especially in developing countries. Pipe sections are usually ordinary shelf products.

If you pay for a professional welder, the price of your device will increase significantly and may even double. You can save costs by investing in all the materials and preparing parts that have been set and welded before delivery. The less the welder has to do the design, adjustment and grinding, the less time is spent on the project, which reduces the hourly cost. This preparatory work is not a particularly great effort, and investment in tools is currently very reasonable. Therefore, you might think that this approach saves a few dollars in the negotiations.

3.3   Value of Time

Unfortunately, for some of us, it is a blessing that we want to learn and achieve instead of making a profit. This also applies to those who work in the construction of a domestic distillery. However, investing a lot in an ethanol project is a good decision, especially for those who are not totally committed to producing large amounts of alcohol. It reduces the amount of these investments (and, therefore, the risk) and also gives you an accurate understanding of the equipment you have never experienced during the purchase.

When you are in the ethanol plant, you must give something of value to your time, even if it is minimal. By using hourly labor costs to collect and process raw materials, maintain the distillery and manage the ethanol product and its billing, you can honestly and accurately calculate what it costs to be independent of the normal distribution network.

3.4   Calculate cost per gallon

It is not difficult to know how much it costs you one liter of ethanol because the price of cooking / heating fuel and raw material remains stable. On a traditional farm, production costs are well-established and independent of the yield per acre and the market value of the crop until the actual profit level is calculated.

The situation is similar to ethanol, although many producers, especially those working in the field of degradation and surplus processing, are not concerned about the value of non-starch or sugar. .

In order for your calculation to be consistent, it makes sense to change the alcohol yield to a standard test measurement, especially if you remove varying percentages of the ethanol test.

I defined earlier in the section how the authorities calculate the tax revenues from the measurement of ethanol; you must use a similar method to determine the value of the fuel. For example, if you make 185 liters of ethanol in one pass and 50 liters of ethanol in one pass, you can conclude that the yield is 140 liters of 100% ethanol. Of course, the actual product is not so clean, but you just have to define a common denominator that you can use in the calculation. The following table shows how these calculations are performed.

Proof Measure Calculation Formula

Number of gallons X Proof strength of product ÷ 100 = Proof-Gallon measure.

Example: Batch One is 50 gallons at 190 proof.

$$50 \text{ X } 190 = 9,500 \div 100 = 95 \text{ Proof-Gallons}$$

To establish a common denominator to evaluate the cost of a series of batches, the runs can be averaged:

Example: Batch One is 100 gallons at 185 proof. Batch Two is 50 gallons at 190 proof.

100 X 185 = 18,500   50 X 190 = 9,500

18,500 ÷ 100 = 185 Proof-Gallons   9,500 ÷ 100 = 95 Proof-Gallons

$$185 + 95 = 280 \div 2 = 140 \text{ gallons at } 100 \text{ Proof}$$

If you prefer to work in percentages, the same result can be obtained using the identical example:

185 proof = 92.5 percent   190 proof = 95 percent

.925 X 100 gallons = 92.5 gallons   .95 X 50 gallons = 47.5 gallons

$$\text{Total} = 140 \text{ gallons at } 100 \text{ Proof}$$

Photo from: R. F. (2009). *Alcohol Fuel A Guide to Making and Using Ethanol as a Renewable Fuel.*

Once established, you can determine the raw material costs of your raw material, calculate transportation costs at your location and reduce the value of each by-product, whether sold or used by you as a grain, also at its fair market value. These include carbon dioxide for bottling and cellulose products that can be used to produce methane or dried in oil.

At this point, you have a net value of the raw material for which it must now estimate the cost of conversion to ethanol. The operating costs associated with this process include the cost of consumables such as enzymes and yeasts, mash, cooking costs, distillation boiler heating, insurance, licensing and financing. They are added to the net quantity of raw materials to obtain the cost of ethanol before depreciation and other miscellaneous expenses, such as electricity required for pumping, maintenance and repairs. Depreciation can be the cost of purchasing leased equipment or machinery that can be extended or reduced over a period of time, typically five years. Labor costs can also be taken into account, although they may change as production increases or decreases.

The total cost of the fuel is then calculated by adding the above adjusted cost to the adjusted cost of ethanol to obtain the net cost.

Calculating Cost per Gallon	
Cost/ Raw feedstock	$137.48
Cost/ Transportation	5.44
Subtotal	$142.92
Credit/ Byproducts	34.80
Net Feedstock Cost	$108.12
Cost/ Enzymes and yeast	5.23
Cost/ Heat source fuel	2.14
Costs/ Fixed	1.66
Pre-adjusted Total	$117.15
Cost/ Depreciation and labor	8.65
Net Cost	$135.80
Yield/ Gallons pure alcohol	76.2
Cost of Alcohol Fuel	$ 1.78

Note: Figures are representative. Fuel, labor and feed-stock costs can vary widely depending upon particular circumstances.

Photo from: R. F. (2009). *Alcohol Fuel A Guide to Making and Using Ethanol as a Renewable Fuel.*

Dividing this number by the number of gallons of pure ethanol (not the actual gallon) gives you the price of a gallon of hard-earned product. Since 2005, the tax credit for small producers has been provided by the adoption of the 2005 Energy Policy Act today, a small candidate manufacturer of ethanol, which by definition produces less than 60 million gallons per year a tax credit of \$10 per gallon for the first 15 million gallons. Some countries may also have other incentives for small producers. The next chapter will examine which is undoubtedly the most important factor in determining the economic and fundamental value of alcohol-based fuels - raw materials and feedstock.

R. F. (2009). *Alcohol Fuel A Guide to Making and Using Ethanol as a Renewable Fuel.* Canada, New York: New Society. Retrieve from: www.newsociety.com

4 RAW MATERIALS

4.0 Introduction

Raw materials that are used in the production of ethanol are the feedstock. In theory, at least most agricultural plants and products can be used as raw materials. Some crops easily produce simple sugars required for the production of alcohol; others are starches and must be divided into their complex form to produce these sugars. Some crops have high yields per acre, but may require special harvesting equipment or other adverse requirements that weaken their attractiveness as a raw material producing alcohol. Useful crops for feed can generally be grown on marginal land, while others are suitable for ethanol production, but are also withdrawn from the market. However, the ethanol production process is similar for all fermentable materials.

The fermentation takes place when the microorganisms (in this case, the yeast) convert the simple sugars into a mixture of liquid mixture in ethyl alcohol and carbon dioxide, in addition to producing heat and an enzyme called adenosine triphosphate. The fermented liquid, called "beer", contains a small percentage of alcohol and, therefore, must undergo a distillation process to increase the percentage needed to produce high quality ethanol. Yeast is a unicellular microorganism, specifically a Saccharomyces fungus, which produces alcohol and carbon dioxide under aerobic conditions.

Yeast makes its work by converting carbohydrates into raw materials as alcohol when these starches and sugars are in a form that yeast can use. Enzymes, catalytic proteins promote the chemical processes necessary for the decomposition of starches and more complex sugars in a usable form, which makes this possible. In general, two types of enzymes are used, alpha-amylase, which initially reduces starches and glucoamylase to complete starch conversion to sugar.

The following chapter explains this complex relationship in more detail. The simplest and most common sugar in the vegetable substance is glucose, a compound containing six carbon atoms, twelve hydrogen atoms and six oxygen atoms, expressed as C6H12O6. This simple sugar is a monosaccharide or fermentable sugar. All crops and all crop residues contain these sugars or their compounds.

Average Alcohol Yield per Weight from Raw Material		
Feedstock Material	Yield in Gal./Cwt.	Yield in Gal./Ton
Wheat	4.25	85.0
Corn, Field	4.20	84.0
Buckwheat	4.17	83.4
Raisins	4.07	81.4
Sorghum grain	3.97	79.5
Rice	3.97	79.5
Barley	3.96	79.2
Dates, dry	3.95	79.0
Rye	3.93	78.6
Prunes, dry	3.60	72.0
Molasses, blackstrap	3.52	70.4
Sorghum cane	3.52	70.4
Oats	3.18	63.6
Figs, dry	2.95	59.0
Soybeans	2.44	48.8
Sweet Potatoes	1.71	34.2
Crabapples	1.29	25.8
Yams	1.36	27.3
Peanuts	1.35	27.0
Potatoes	1.14	22.9
Sugar Beets	1.10	22.1
Figs, fresh	1.05	21.0
J. Artichoke	1.00	20.0
Citrus Waste	0.83	16.6
Pineapples	0.78	15.6
Cranberries	0.78	15.6
Sugar Cane	0.76	15.2
Grapes	0.75	15.1
Apples	0.72	14.4
Apricots	0.68	13.6
Pears	0.57	11.5
Peaches	0.57	11.5
Plums	0.54	10.9
Pumpkins	0.49	9.8
Carrots	0.49	9.8

Note: Alcohol yields at 190 proof. Yields compiled from Goosen's EtOH Fuel Book (1980) and USDA sources. Short Hundredweight (Cwt.) equals 100 pounds US. Short Ton equals 2,000 pounds US.

Average Alcohol Yield Per Acre from Raw Material	
Feedstock Material	Yield in Gal./Acre
Sugar Beets	287 - 412
Sugar Cane	268 - 555
Corn, Field	214 - 390
J. Artichoke	180 - 613
Potatoes	178 - 299
Sweet Potatoes	141 - 190
Apples	140
Dates, dry	126
Carrots	121
Raisins	102
Yams	94
Grapes	90
Peaches	84
Prunes, dry	83
Pineapples	78
Pumpkins	78
Cranberries	70
Rice	66 - 175
Soybeans	58
Pears	49
Barley	48 - 83
Molasses, blackstrap	45
Apricots	41
Peanuts	40
Oats	36 - 57
Sorghum grain	35 - 125
Buckwheat	34
Wheat	33 - 79
Figs, fresh	31
Figs, dry	29
Sorghum cane	26 - 500*
Rye	24 - 54
Plums	22
Citrus Waste	N/A
Whey	N/A
Crabapples	N/A

nearest whole number. Yields based on multiple sources including USDA Misc. Pub. 327, December 1938 and USDA Agricultural Statistics, 1978 and 2003.
* Lipinski, E.S., "Fuels from Sugar Crops." 1979

4.1   Types of Feedstock

Depending on the composition, the raw material is classified into three different categories: which are sugar crops, starch crops and cellulose crops, such as forage and residue. With current technology and the rise in the price of agricultural products and energy, traditional models of ethanol production, that is, the use of corn starch as raw material and natural gas as a boiler fuel must be reevaluated with a critical eye.

A holistic approach can be economical and sustainable, and much healthier for the environment. For example, in the United States, and particularly in developing countries, ethanol producers consider non-traditional crops, agricultural production, waste and food industry waste as a possible raw material. Small businesses are particularly sensitive to these alternatives because they are more flexible than large commercial or industrial interests and are more useful for experimentation. The following is a large sample of conventional and unconventional stocks that can produce ethanol.

Sugar Crops

Most sugar crops are easily recognizable by their names. Sugar beet, sugar cane and sweet sorghum are common examples. Other crops, such as Jerusalem artichokes and fruit plants (apples, peaches, and pears, to name a few), also have high levels of fermentable mono saccharide and may be potential candidates for ethanol production.

The simple six carbs found in most of these crops occur singly or in pairs. This non-complex structure is easily fermentable because it is sufficient for the raw material to be crushed or ground to release the sugar in a form that allows the yeast to work. The material used in manufacturing is widely available and widely used in agriculture and includes grinding machines, hammer mills and extraction machines.

However, the ease with which these sugars are converted is a disadvantage because these crops are susceptible to contamination or spoilage. The microorganisms that pollutes and responsible for this spoilage develop in sugars that are very humid and rich in moisture. In the presence of air, the sugars are converted to acetic acid and carbon dioxide.

This problem can be solved by using the raw material directly after harvesting or by evaporating it to remove the moisture and store it safely. From an economic point of view, drying consumes a lot of energy and leads to additional equipment costs. Solar dehydrators can reduce operating costs with conventional evaporators, but will still require a capital investment.

SUGAR BEETS

Sugar beets can tolerate a wide range of soils and climatic conditions and are widespread. They are particularly suitable for colder climates, where other crops are not so good. The treatment requirements are low and with mechanical milling followed by compression as a selection procedure to extract the sugary juice. Beets contain about 16% sugar and rise to 85 ° F in a short time between 8 and 24 hours. The beets can also be milled with hammer and mixed with water to form slurry. However, it must be cooked for at least one hour before the yeast is introduced. As a root crop, it requires rotation with species other than beets to control parasitic nematodes. The ethanol yield is relatively high; in some cases, there are up to 750 gallons per acre.1 22 liters per product are common.

The sugar beet also has the advantage that it is stored well in stacks of up to four months and offers valuable by-products. Post-treatment of the residual mass can be used as food for distillates, and beets provide the fertilizer with organic matter.

SUGAR CANE

In the United States, this crop is usually limited to four states (Texas, Florida, Louisiana and Hawaii) due to climatic constraints, but in Brazil as in the ethanol industry which has been built on it. It is one of the largest ethanol concentrations per acre, ten times more than the corn in the right conditions, and also offers a remarkable return on the remaining biomass (bagasse) that can be used for boiler fuel and cogeneration after mechanical or manual processing.

In addition, diet-based digestion, which has been retained from distillation after a food rack, is used as a liquid fertilizer for next season's crops. Sugarcane is a perennial crop; it is planted once and then harvest for five or ten seasons. Much of it is made from sugar cane because of its alcoholic potential, and then it is considered a great example of the ethanol economy. However, in reality sugar cane cannot flourish at temperatures below 45 ° C and passes temperatures that exceed freezing, so its potential for small scale ethanol production is limited in most US and Canada countries.

Although certain crops can be planted and harvested annually, a longer growing season is required between late spring and late frost. Although this increases the cultivation potential, it is also more labor intensive. In Brazil, the traditional cultivation of sugar cane is not suitable for the environment, as the harvest is first burned to remove the leaves and expose the stems of the plant. After that, the laborer army cuts its arms and stack for processing.

Recently, employees have been working to manually remove leaves (with contain biomass) manually (use of hand) this serves to create jobs and reduce particulate pollution. In well-activated plantations, sugar cane and leaves are collected with the harvester though is a time efficient process but it is costly. It is intended that manpower or capital investment is necessary for sugar cane propagation, as is the case with other crops. However, cane juice regularly receives about 750 liters per acre, and the value of the seeds and leaves, which are high at half the weight of the plant, is a fuel and a conversion of cellulose.

The sugar cane is usually crushed in a drum crusher or roll mill which squeezes the juice, collects and sends the stalks out to dry. Crushers are made of any size and can sometimes be found as salvage. Another method for extracting sugar is known as diffusion, in which the thin parts of the tube are immersed in the solution and heated to extract sugars from the substrate material of the solution.

The enriched solution is passed through several steps and the additional sugar is extracted from the new substrate stock at each step. Once the sugar supply is exhausted, it is removed for use as skimmed or forage, and new pipe sections are added. A new substrate is introduced into this rotating substrate until the tube is fed.

Sugar cane is stored only in a few months in stalk form and needs to be used immediately after juiced. Fermentation is fast and will be completed in less than 24 hours. No cooking needed.

SWEET SORGHUM (Sorghum bicolor)

In North America, two types of sorghum are grown: grain sorghum, called sorghum bicolor, and stalk-like cane, called sorgo or sweet sorghum. Sugarcane cultivation is a viable candidate for ethanol production in soft areas. Good ethanol content per acre and a high degree of tolerance to different climatic and soil environments will give sweet sorghum a welcome flexibility as a raw material (feedstock). Some varieties produce both grain (milo) and sugar, which increases their value as a raw material (feedstock).

During treatment, the plant can be fully harvested and stored or processed to remove sugars after harvest. It can also be mechanically hacked and green fodder stored in the silage. The high protein content of its leaves and fibers make it an excellent food. In addition, like sugar cane, fresh sorghum residues can be burned as a heat source for ethanol treatment processing. Sorghum has high sugar content and can produce two growth cycles for a warmer air conditioner.

The yield of alcohol may vary from 400 to 600 gallons per acre. To extract the beverage from the sugar cane, the short parts can be crushed in the roll mill, but the fiber must be separated. On a small scale, crushing and fermenting the entire product can be more costly. Grain (milo) sorghum can provide significant yield in some varieties. After harvesting, the grains must be ground and the liquid mixture must be prepared, then heated and hydrolyzed or disintegrated with other grains. The following section on starch crops describes this process in more detail.

MOLASSES

The molasses is not a crop itself, but it comes from sugar cane or sugar beet juice and what remains when the juice crystals are removed to make a table and culinary sugar. Molasses is used as a supplementary feed for livestock and is one of the major ethanol production sources for commercial alcohol before the Second World War.

Blackstrap molasses, a concentrated byproduct of sugar crystal production, is almost ready for ethanol production. The sugar content of molasses varies depending on the source, but varies between 35 and 50 %. Thick syrup can be a potential raw material (feedstock) when it is available, but the cost can be high and generally vary on the open market. You can have an idea of whether molasses is an economically viable resource when you look at the ethanol yield: 70 liters, 95% pure alcohol per ton. At a market price of $ 100 per ton, this is unlikely. However, if the cost decreases to $ 50, there may be enough room to cover the costs on a small scale.

The molasses requires heating and dilution of hot water to obtain the purge or mash solution. Pumping molasses without heating can be a challenge without preheating the pump. The mash diluted with 15% sugar may require some yeast. After the original inoculation of the yeast, the fermentation must be completed within 48 hours.

Jerusalem Artichokes
Jerusalem artichoke has shown great potential as an alternative sugar crop. This harvest comes from the

sunflower family and comes from North America and is adapted to the northern climate. Like sugar beet, Jerusalem artichoke produces sugar in its superior growth and stores it in roots and tubers. It can grow on different soil and does not require soil fertility. Jerusalem artichoke is a perennial plant; the remaining small tubers in the field produce next season's crop, so there is no need to plow or sow.

Although Jerusalem artichoke has traditionally been grown for tubers, there is also an alternative to tub collection. It has been found that most of the sugar produced in the leaves does not enter the tuber until the plant has almost reached the end of its useful life. Thus, it may be possible to harvest for Jerusalem artichoke when the sugar content of the stem reaches its maximum value, thus avoiding the harvest of tubers. In this case, the equipment and harvesting procedures are essentially the same as for harvesting fresh grapes or corn.

Fodder Beets

Another promising sugar factory being developed in New Zealand is the fodder beet. Fodder beet is a high-yield forage crop obtained by crossing two other types of beet, sugar beet and mangolds. Sugar beets seem to be similar in most agronomic aspect. The attraction of this crop is its higher yield of fermentable sugars per acre compared to sugar beets and their relatively high resistance to the loss of fermentable sugars during storage. The cultivation of fodder beet is also less demanding than sugar beets.

Fruit Crops

Fruit plants such as grapes, apricots, peaches and pears are another commodity in the category of sugar plants. In general, fruit plants such as grapes are used as raw material (feedstock) for wine production. It is unlikely that these plants will be used as raw materials for fuel ethanol production because of their immediate market value for human consumption. However, by-products derived from the processing of fruit are likely to be used as raw materials as fermentation is a cost-effective method to reduce the potential environmental impact of untreated waste containing untreated sugar.

Starch Crops

It is more difficult to treat starch crops than sugar plants because their sugar units combine into Long Branch chains. These complex chains of starch must be divided into single units (or paired) of six carbons, so that the yeast can convert sugars into ethanol. Fortunately, the starch conversion process is a fairly simple process that uses heat, enzymes or acids.

Enzymatic hydrolysis is a process in which enzymes and water disintegrate the starch after it has been made available by grinding or milling. The biggest disadvantage of using starch plants for the production of ethanol is the additional energy, equipment and time or labor required to complete this decomposition process.

In particular, energy is an economic aspect that requires careful analysis because of its volatile and unpredictable costs. However, small scale producers have more flexibility than large users and can work more easily with alternative heat sources such as wood, crop residues and even solar heat to meet their needs. In addition, advances in enzyme research have resulted in strains that function well between 85 ° C and 104 ° F with the conventional process, instead of the usual 185 ° F to 219 ° F. advantage that there is no sugar culture and that it has a storage potential.

Factors that prevent the initial conversion prevent the starches from being easily stored. When the grains are dried, it is very difficult for the microorganisms to work with the starch and proteins in culture. Long-term storage is a definite advantage for ethanol production because it provides flexibility and opens up markets for additional raw materials.

Cellulosic Crops

Discussions on fuel supply have intensified research into the use of cellulosic materials as a source of ethanol. The use of cellulose instead of starch or sugar is an obvious problem of distributing food for human consumption for energy purposes. At the same time, it encourages the cultivation of marginal lands and less demanding plants in water and fertilizers, thus saving these resources for the efficient production of food.

The stems and leaves of the typical starch and sugar crops are mainly cellulose. Cellulose contains sugar, but individual sugar units bind long chains with chemical bonds much stronger than starch. Since cellulose must be broken down into simple sugar components to make the yeast work, conversion is problematic due to the additional steps and associated costs.

Cellulose is wrapped in lignin, a complex compound that is part of the cell wall and gives the wood its typical stiffness and strength. Lignin is very resistant to enzymatic and acid hydrolysis. Biotechnology companies, such as the Canadian Iogen, and universities such as Perdue and Penn State, are establishing demonstration facilities using advanced technologies, but the cost of converting liquefied cellulose into fermentable sugars is still far superior to the means and scale of a small-scale production facility. As the cost of conventional oil fuels increases, the conversion of cellulose and the development of its associated enzymes will of course become more attractive and ultimately economically viable.

In order to get an idea of how difficult it is to solve this problem, the research done in the late 1970s and the work that followed in the early 1980s almost unanimously assured that the cellulose conversion is "immediate" or "few years" but it is unlikely that small, alcohol-consuming efforts will be justified by new technologies. "" Wellbeing "before they become economically viable at the local level. As fuel prices rise, potential small-scale alcohol producers must be prepared to separate wheat from hell in order to speak and openly deal with extraordinary claims, particularly in terms of investment.

Crop Residue

Wheat straw, Stover (dry stalks) and wood used to make ethanol (such as the fast-growing hybrid poplar) are mainly cellulosic materials. Because of the complexity of the bonds in cellulose, wood is not economically viable for small-scale ethanol production except as fuel for heating equipment. In the last conversion of cellulose to cellulose conversion described in the following chapter, a general description of the ethanol process from cellulose is given.

Waste, Surplus and More

A small farmer with the curiosity and time to experiment can find a gold mine for wild growth and agricultural or commercial waste. Many plants are being investigated to determine the viability of ethanol production and the use of crops and surplus, abandoned or even contaminated food is not new. However, the rising costs of petroleum-based fuels have sparked interest in these "odd balls", and some may be particularly suited to small-scale production under the right conditions.

The Dynamics of Yield

After reading this chapter, it is difficult to calculate the solids content of the raw material (feedstock). It may not be as important if you only test renewable fuels or make ethanol for your vehicles. However, if you want to share ethanol, you must first determine the potential yield production. I can tell you that the two most important variables of a beginner are the performance and yield of ethanol from a particular raw material (feedstock).

From an academic perspective, crop production is no more complicated than access to USDA extension or data services. These numbers depend so much on the local variables and the environment variables that the best you can expect is the average closest to what you get. Since many potential fuel producers buy, sell or market their feedstock (raw materials), they never participate in agricultural production, so the problem is less serious. The amount of ethanol that you get from your inventory also depends on almost invariable variables.

Processing methods, cleaning, selection of enzymes and yeasts, temperature management, environmental management and equipment quality, as well as their experience are taken into account. The best advice I can give you is to keep accurate information during production, study and learn from your mistakes.

When looking at fuel oil related to agricultural alcohol, there is a significant correlation between yield per acre and yield in ton, which can be misleading if, for example, . It is in the top third of alcohol yield on a per-ton basis, but right near the bottom in yield per acre. The carbohydrate content or the energy density of the food explains that, but it is an important factor. If you grow a feedstock (raw material) for ethanol production with a limited number of acres, you want a manageable crop with a relatively high yield per acre.

On the other hand, if you buy feedstock (raw material) elsewhere or look for spoils, the return per ton is much higher (considering the costs, of course) because you do not really care about the order of growth. The price of energy is also a very important factor in determining whether the ethanol it produces is worth it. The dynamics of inverted energy (EROEI) surpasses the kitchen and yet its efficiency. If you do not consider the cost of transporting the raw material, prepare it if necessary for storage, handle it efficiently and store it without loss, rejection or disappointment.

Always remember that sometimes the ethanol yield is only part of the puzzle. The value of by-products, such as by-products of feed additives or the manufacturing or distillation process can significantly offset the costs and make the yield less important. The raw material hulls or shells can be a commodity for the industry you do not know yet. Carbon dioxide is easy to use and has commercial value. Enzymes and cultured yeasts can be a value-added product; the treatment of hot water can be recycled to another part of your operation. Stalks and hulls can be burned or converted into palletized boiler fuel.

As part of the cooperative operation, it may be more convenient to install distillation units in a food factory than to transport the raw material to a separate alcohol factory. In Brazil, hundreds of independent sugar cane factories set up small ethanol factories using juice extraction, which uses the same machinery and process as sugar or alcohol is produce or not. In the United States, other crop processing uses the same pattern.

R. F. (2009). *Alcohol Fuel A Guide to Making and Using Ethanol as a Renewable Fuel.* Canada, New York: New Society. **Retrieve** from: www.newsociety.com

5 SUGARS AND FERMENTATION

This is one of the most important chapters in this book. It covers the "culinary" aspects of alcohol production: preparation, cooking, fermentation and all the little details that make the recipe a profitable success. It's not a chapter on easy assimilation, and it's still a long time, but after a little test, it makes sense. In order to maintain a consistent and sequential data flow, I have used both starch and sugar techniques, although they are two different types of crop. Finally, the fermentation process is essentially the same, but I point out to cases where subtleties occur.

5.1 Basic Starch Technique

Starch crops differ from sugar crops when using glucose, monosaccharide or simple sugar. Although a mature sugar plant, such as sorghum cane, stores glucose (as well as cereal starch), the starch crop converts the remaining glucose structure after the construction of the cellulose and lignin (grains) or tubers (tubers). Starch is only glucose polymerized units, molecules that are linked by long chains.

The chains may be hydrolyzed or disrupted into individual glucose units, in stages, on the one hand, to dextrin, a short-chain glucose intermediate, then disaccharides (maltose, sucrose, lactose and mono-glucose), two simple sugars and a molecule that the yeast can digest. The starch can be hydrolyzed in various ways: with hot water, pressures, acids, and the use of fungal or bacterial enzymes.

Some of these alternatives are too expensive in the home or farm scale and too complicated to be practical. Therefore, small scale hydrolysis of starch is normally limited to hot water cooking and commercial enzyme or malt treatments. The technique can be divided into five basic steps:

1. Milling
2. Suspension
3. Hydration
4. Post-liquefaction and

5. Conversion

The sixth step, the fermentation, takes place when simple sugars are available, and this step is essentially the same as using starches or sugar feeders. Various ways of describing this process may be a separate hydrolysis step and a liquefaction step, but the technique remains essentially the same.

Corn and other crops containing zinc starch require a two-step process in the slurring step, including heating and boiling; Pre-Malting, which brings with it a small amount of alpha-amylase enzyme, facilitates mixing to avoid accumulation and excessive thickening at higher temperatures. Softer starches such as potatoes and rye can be treated at slightly lower temperatures, and the steps can be combined with the use of certain enzymes.

Step 1: Milling
The reduction of starch molecules begins with the physical grinding of seeds or grains into pieces that are small enough to expose the granules to the water and be available for the enzymes. Soaking, grinding, cracking or rolling the grain is not enough to expose

the starch; it must be hammer mill or grinding mill with a medium or fine meal.

The finest flour less than 1/8 inch is used in commercial operations where grain harvesting is done with sophisticated equipment. However, if you want to remove the grains from the distiller after fermentation, it is difficult if the material is ground into fine particles. If you do not intend to restore the distiller's grains, the performance of the alcohol will increase to a point when you grind it to finer grains.

Most rural areas have a community mill or a feed store where a small fee can be charge for grinding your grain. However, if you take ethanol production seriously, you should probably invest in a small hammer mill or grinder. Residual material, such as stems or bottles, must be filtered from the raw material before grinding. They have no effect on the mash, except for the addition of large substances that cannot be converted to glucose by the techniques described here.

Step 2: Slurring

Hot water is sufficient to swell the starch granules to decompose and release the amylose (soluble inner portion of the starch cell) known as gelatinization. The amount of water required per bushel or the weight of grain varies with the type of crop and the method used. The trick there is not to add too much to avoid diluting the sugar, ideally in the range of 20%. It is possible that water will not be used immediately (depending on the recipe), but it should always be in the pH range of 5.5 to 7.0, which contains no chemicals and does not contain enough to feed to livestock , "Sweet water" after using distillery products.

Corn weighs 56 kg and 65-72% starch. The suspension and hydrolysis of ground foods requires about 28 gallons, possibly more. You do not want to compromise on fast gelatinization because the viscosity of the wort (mix of grain and water before adding yeast) is so high that mixing becomes difficult. This is avoided by increasing the grain mix before the temperatures of 155 ° F are transferred to the gelatinization range, but it can be significantly higher at higher levels of amylose.

Start with 20 gallons of water per bushel, slowly add the grain and bring the temperature to 155 ° F (155 °

F). If you used some of the used grease in a previous use, the temperature is initially warm. It is important to make a small agitation as soon as the grain is introduced, as this helps to maintain a consistent consistency of the must and avoid burn marks and hot spots.

For other cereals such as rye, barley or rice, the water content is very similar and the carbohydrate content is also close to that of corn. Their gelatinization temperatures vary, but are generally lower than that of corn. At this stage pre-malting treatment may be performed on corn and other similar hard starches. Alpha-amylase is added in a modest amount, about 1/2 ounce per bushel of cereal, and the temperature is raised to 190 ° F or higher while the agitation is maintained. At this stage, do not allow temperatures below 180 ° F as the starch may decrease to a thick mass despite the enzymes. In a few minutes, the spice is refined and gives a strong, almost bitter aroma.

Continue heating until boiling; Corn does not hydrolyze completely if left inactive for at least 45 minutes. The high temperatures in this cooking process destroy the initial alpha-amylase dose, but they have already reached their target. In later stages of liquefaction, a follow-up dose is used.

Step 3:   Liquefaction

Hydration (conversion of starch to water-soluble semi-complex oxide) is necessary to prevent the suspension from gelling to a massive viscous mass, just as in the process of preheating hard starch. The combination of heat enzymes and alpha-amylase, whether commercial or malt, begins to break down the starch into dextrin. At this point, it is important that temperature, movement and pH are to minimize liquefaction times, not to reduce treatment costs, as well as to reduce the risk of bacterial infection. The mixing time of 45 minutes agitation should be sufficient and temperatures between 180 and 190 ° F favor an effective reaction.

PH values vary throughout the process, but should be between 5.0 and 6.5 (commercial enzyme manufacturers determine the temperature and pH of their products). Alpha-amylase enzymes are used at about ¾ ounces per bushel or at the recommended dry weight levels of carbohydrates in animal feedstock.

If you use different raw materials or cuttings and surpluses, it may be difficult to determine the exact proportion of carbohydrates in the raw material. In these cases, it would be advantageous to run small scale tests by mixing a known amount of dried raw material with a known amount of water (8.33 pounds per gallon) to obtain a two gallon test work. The oil can be boiled slowly for about one hour or so, and then allowed to cool to about 190 ° F before the pH is adjusted.

When done, you can divide the bottle into five or six equal volumes in clean and separate containers. Calculate the amount of yeast needed for a small volume of wort, and gradually add one to the control and the other doses by storing the measurements for each one. Store the containers at 190 ° F in a container filled with water and hold the samples with stirring for several times for an hour. At this stage, you can cool the samples and run an iodine test on each of them to determine the most effective dose of starches in sugar. The sucrose sugar test helps to indicate percentages of sugar processing and raw materials.

Step 4: Post-Liquefaction

The progress of enzyme technology and the benefits of industrial technology have made this step less annoying for some. However, for small operators, post-liquefaction may be crucial for the total dextrinize of the remaining starches that are in the wort. This can be done by simply adding a second dose of alpha-amylase enzymes to hydrolyze this starch. The temperature is maintained or reduced to the optimum range of the enzyme and, if necessary, the remaining excess water is added and the pH is checked before adding the enzyme at a rate of ¾ ounces per bushel which can vary.

Post-liquefaction may take 15-30 minutes. It should be noted that softer starches, such as potatoes and barley, are easier and cheaper to handle because the liquefaction and cooking process can be simplified. This is because soft starches do not need to be cooked (as with corn), but must boil at 200 ° F to liquefy. Commercial alpha-amylases enzymes survive for more than an hour at these temperatures and continue to operate in their preferred range of 180 to 190 ° F. With continuous mixing and tight temperature and pH adjustment, a lower heating time results in energy savings.

Step 5: Conversion

The conversion step is also called saccharification, and starts when most of the starch is converted to dextrin. (You must understand that a 100% conversion is probably an unrealistic small target - a very small portion of the starch is likely to be retained because it is not completely gelatinized). He said simple sugars. It is now the case that the dextrin sugars are converted into glucose, fructose, sucrose and other simple sugars mentioned at the beginning of this section.

Another task of the enzyme, glucoamylase (or barley malt beta-amylase), does this. Commercial glucoamylase enzymes are likely to be the transformation of starch and sugars are more effective in the task simply because they involve a wider range of dextrin, but their cost can replace some of that efficiency for those who want to maintain at least a sustainable yield in supplies, Take at least ¾ ounce per bushel of grain and much more for some commercial enzymes. (The manufacturer specifies the requirements in his database.)

The first step in the conversion is to lower the temperature of the wort. This can be accomplished by pumping cold water through cooling coils or simply adding water by adding the mixture to about 140 ° C (140 ° C). Too high dilution affects the percentage of next sugar (which is about 16-22% at the end of the process). However, check it again before you start fermenting.

In general, the temperature in this stage should not rise above 150 ° F or fall below 125 ° F, but there are enzymes that remain below 105 ° F. along with development of yeast in the cold cooking process which described in the section on recipes for planting starch. Previous sugars at low temperature with Biocon enzymes (USA) have shown that, after inoculation with yeast, probably up to eight hours of fermentation time could be attributed to the shortage of fermentable sugars.

After adjusting the temperature, you may need to test and adjust the ph. Do this after adding water, since acidity or alkalinity will affect the pH of the wort. The

PH must be approximately 4.5 or at least lower than the values observed in the previous steps. Some commercial enzymes and beta-amylase may require a slightly lower pH, but the manufacturer's recommendations and experience guide it. Remember that lowering the pH means adding sulfuric acid or phosphoric acid. Therefore, be prepared to take proper precautions.

At this point, the enzymes can be introduced into the wort. Depending on the enzymes used, the conversion of sugars takes 30 minutes and several hours. The wort should slowly agitate. To check your progress, perform the iodine test as described in the "Tests" section of this chapter. The goal is clear reading or a little pink tint. If the test shows traces of blue or purple, the force is not completely converted, and you need to spend more time on conversion. If after a few hours the controlled sample does not come out clearly, you have to live by the fact that the percentage of the mass cannot be used for fermentation.

Prepare the finished ink at a temperature of about 90 ° F or use the yeast for the last and last fermentation. The use of cooling coils saves the addition of cold water, which can change the pH level. Wort must measure the sugar content in the range 16-22%, measured by a saccharometer or refractometer. Once the target is reached, you can stop the movement or agitation and put the wort in the fermentation tank or vessel.

5.2   FERMENTATION

The fermentation is a natural conclusion that I brought you through, but it's really the same process, because when the raw material or the substrate is converted to simple sugars, the fermentation process is very similar, regardless of whether it had sugar or starch crop used. In the section "The Importance of Yeast", I explain how these microorganisms go through various stages within 24 to 72 hours to break down these sugars and produce carbon dioxide and alcohol. I'll just say it's a chemical process where the yeast cell and its enzymes absorb glucose, fructose or mannose, and the enzymes and coenzymes carry the sugar molecule through 12 intermediate changes.

At the beginning of the 20th century, the biochemist Gustav Georg Embden conducted research and Otto Meyerhof characterized this process, which is known as the classic Embden-Meyerhof cycle. The fermentation takes place in a closed vessel, ideally a process (usually a heat exchange coil or jacket) for heating and cooling the fermentation mixture, where oxygen is introduced into the initial stages of the process, and a valve connected to allow CO_2 to escape, also an agitator to stir if available. The container also requires a fill and drain system, access port or lid and some means of cleaning. See Chapter 7 for more information about this device. Fortunately, a small scale producer has plenty of creative flexibility to design the devices that work best for them.

Once the wort is made and the ideal temperature range is set at 85 ° F to 90 ° F, it is transferred to a fermentation vessel (in some environments it is boiled and kept in the same container). At this point, you should introduce the yeast. There is a method used for "pitching", which I described in "making a yeast starter" section, but you can also add yeast into the wort. The difference is actually how long the fermentation begins and becomes really energetic. With a higher yeast dose and a high volume of pitching solution, the reaction can start in a few hours. Otherwise, it may take half a day or more to appear active.

Even if you have just added yeast to the wort, you have to give organisms an advantage. Take three to five liters of fresh, warm water (ideally 90 ° F to 100 ° F) in a convenient container and add three or four ounces of malt (or germinating wheat). Then, dry yeast is added at a rate of about ½ pound per 100 liters of liquid. This corresponds to approximately 2½ pounds in a 500-gallon batch, but thicker or softer pulpy mashes and those with insoluble sugar will improve at higher slaughter rates, up to twice the yeast.

Immediately the yeast is introduced, you can shake the mixture vigorously and let it work for about 20 minutes. Adding air to the mix by splashing or sometimes mixing will help spread the yeast. Before continuing, I want to emphasize that the water must be clean and free of any cleaning agent because it has an immediate and damaging effect on yeast cells. If you want to try it out, you can boil the water sample and let it stay uncovered so that the chemicals evaporate and disappear. Several distillers will suffer unnecessarily over something preventable, such as the use of chlorinated municipal waters to filter air pollutants in advance. After the first use, you can pour a small sample into the yeast fermentation vessel to multiply.

The introduction of free oxygen into the mash is a key factor in the development of a rapid ferment of the enzyme at an early stage. Oxygen promotes growth and reproduction. If you do not have access to a tank containing welding supplies, the compressed air will be sufficient. Bubbles can be injected through a micro-bubble tube that can be made by drilling a portion of the plastic tube with 1/16 inch holes. It is attached to an oxygen tube and immersed in the bottom of the fermentation tank. The injection should begin with the initial fusion for 15 to 30 minutes, after which the yeast can withstand up to 24 hours to absorb and reproduce oxygen.

When the yeast enters the anaerobic phase, a significant amount of carbon dioxide is removed from the fermentation lock and an airlock is connected to the channel at the top of the closed container. Bubbles indicate that the fermentation is in progress and heatis also generated. With thick mashes, the mixture may need to be mixed slowly to ensure that all the yeast comes into contact with the nutrient fluid. In addition, mixing or agitation helps break the solid cap, which tends to grow on the surface of the liquid that can actually block the yeast supply.

Regardless of whether you use pump agitation or blade designs, it is important that the outside air does not get into the tank during the process. Otherwise, it may be contaminated. After 24 to 36 hours - in some cases up to 72 hours - the bubbling stops and the fermentation site smells of alcohol. Any floating solids will likely sink into the bottom of the tank. At this point, the mash is immediately distilled to prevent the growth of bacteria that can acidify the whole batch to vinegar.

5.3   The Significance of Yeast

Yeast (unicellular microorganisms that transform simple sugars into alcohol) is a much more complicated and arduous organism than you can imagine. There are dozens of yeasts species, but only a handful is valuable to the alcohol fuel manufacturer. To be a good candidate, the yeast strain must be able to ferment quickly, tolerate alcohol, be economically practical and convert the sugar contained in the raw material used. Several products of dry yeast or Saccharomyces cerevisiae are commonly used for the production of industrial alcohol.

Distiller's yeast has a higher tolerance of alcohol, and special yeast used in strictly controlled environments can support more than 20 percent alcohol. So, what happens when the yeast is fed to the source? It is not surprising that they tend to eat and multiply. Of course, things are more complex, so we analyze the process in more detail. There are two phases of life for yeast.

The first stage of life of yeast is the aerobic phase where the environment is full of oxygen. Yeast absorbs oxygen, consumes sugar, in large quantities and releases carbon dioxide. At this stage, very little ethanol is produced, but a large amount of carbon dioxide is produced. The metabolism also generates considerable heat sufficient to warm the environment outside the production level later on.

When the oxygen content is used up, the process becomes an anaerobic stage (second stage of life of yeast) where the yeast consumes the remaining sugar while produce ethanol and emits carbon dioxide. As sugar is metabolized, more heat is generated. About half of the sugar is used for carbon dioxide production, and almost everything else is used for alcohol production. In the production of co-products such as acids, adenine phosphates and fusel oils, a small amount, probably less than 5 percent, is lost.

Yeast Cycle

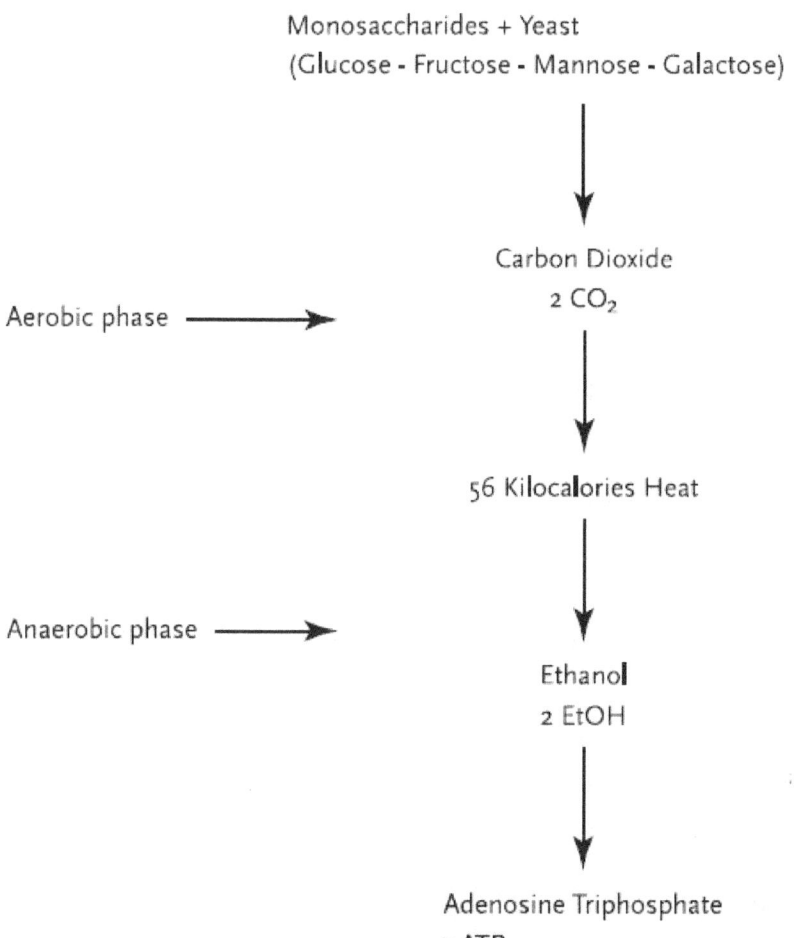

Monosaccharides + Yeast
(Glucose - Fructose - Mannose - Galactose)

↓

Carbon Dioxide
$2 CO_2$

Aerobic phase ⟶

↓

56 Kilocalories Heat

Anaerobic phase ⟶

↓

Ethanol
2 EtOH

↓

Adenosine Triphosphate
2 ATP

Photo from: R. F. (2009). *Alcohol Fuel A Guide to Making and Using Ethanol as a Renewable Fuel.*

5.4   The Role of Enzymes

When foods are metabolized, they are synthesized into complex elements that are converted into simple elements that are then transformed into usable energy. Enzymes are living cell-producing catalytic proteins capable of triggering these normal biochemical reactions without being altered or being destroyed. Enzymes increase the number of chemical reactions and can do so in the form of proteins, whether they come from micro-organisms, plants or animals.

Although enzymes are catalysts, they are also biological components and can be rendered ineffective by extreme heat, acidity or alkalinity. In addition, there are several types of enzymes, each of which is very specific to its function. Unique enzymes have been developed not only for the ethanol industry but also for the manufacture of textiles, wood products, detergents, foods, and many other products.

At a time, the enzymatic reactions are carried out by the molecule. In the hydrolysis process, the enzyme at a particular location of the enzyme is combined with a substrate or raw material (feedstock). When a match is made, the chemical reaction breaks down the substrate molecule into smaller components. The enzyme then dissolves from the substrate, returns the atoms lost in the chemical reaction and repeats the process with the second substrate molecule.

Enzymes work very quickly under the right conditions when the activity rate is between 1 thousand and 1 million. However, these conditions have a clear effect on the efficiency of the enzyme.

5.5   Ethanol's Key Enzymes

Grains and starch-based raw materials mostly use a group of amylase enzymes. These are particularly suitable for the hydrolysis of starches to various types of saccharides, three of which are of optimum importance are alpha-amylase, beta-amylase, and glucoamylase (also called amyloglucosidase).

Alpha-amylase is found in an animal, plant, fungal and bacterial sources. It is present in all starch products and is formed during the grain-malting of a process. Its main objective is to liquefy the starch granules during the cooking process and decompose them into dextrin, then maltose and oligosaccharides if the product can work for long enough. Some bacterial alpha-amylases are designed to operate at 190 ° F or higher for cooking purpose. Beta-amylase is another sugary enzyme found in microorganisms and in plant life. It works with alpha-amylase converting dextrin into maltose, but it cannot avoid glucose branch points (binding points) and bonds, so it leaves part of the starch in the form of dextrin.

Glucoamylase comes from a strain of fungi. It is important because it has the ability to hydrolyze both branches and linkages of starch and thus they can completely convert starch into fermentable glucose. Its main deficiency is that it operates slowly because it hydrolyzes a unit as it moves.

During the fermentation process, when the yeast is added to the saccharified mash, the yeast also produces its enzyme. The enzymes of the additional cells produce cells that make them and maltase and other sucrose. Maltose hydrolyzes maltose to glucose and sucrase distributes sucrose to the fructose unit and glucose unit. Intracellular enzymes ensure cellular metabolism itself; Yeast core takes glucose and ferments as described in the chapter "Meaning of yeast" known as Embden-Meyerhof cycle

5.6 Making Barley Malt: The Home-Grown Enzyme

Malt is a barley grain that deliberately germinated for its enzymes. Although other grains can germinate so easily, barley is chosen as the grain because it produces a good amount of alpha and beta amylases that are used in the mashing process.

This question may arise from "why would barley be in the business of producing enzymes"? The answer to this question lies in the natural cycle of the crop. When the dormant seed is stimulated by the appropriate amount of heat, moisture, and air, it begins to germinate and undergo several changes.

There is germination of stalk and rootlet; it is not so obvious that enzymes formed will eventually break down the cell wall between the embryo and the endosperm, and in the process where the seeds are extracted from the food through the stem and roots starch are converted to sugar and back again. When the germination process is stopped at the peak of enzyme production, the seeds can be dried and stored, or used immediately for their enzymes.

Barley is not a replacement for commercial enzymes. It has good and bad points for everything. This includes time and works as well as moderate investments in seed acquiring, germination and storage processing. The quality may not be consistent, and organic enzymes are unlikely to be as effective as the enzymes produced.

But there are also important benefits: when you have achieved and gained experience, you have found a truly profitable enzyme source, which is completely out of reach of commercial suppliers. You will have an independent supply that is controlled solely by you. If quality is not always the goal, it's easy to use more mushrooms to make up for the deficit and every barley you use, increases overall yield because it's a source of starch itself.

Using Barley Malt

The usual recipe for barley is to use 10 to 12 weight percent malt in relation to the feedstock grain. Since malt contains alpha-amylase and beta-amylase (glucose amylase maltose version), the total volume must be divided into two parts to allow suitable steps to be used in the fermentation process. There are different conditions at each stage to prevent the action of two amylase enzymes at the same time, although both enzymes are still in the malt. In the liquefaction stage, malt-alpha-amylase replaces the commercial alpha-amylase quite efficiently, although its pH and temperature range are not so wide.

The malt enzyme produces better results at a pH between 5.5 and 6.0 and at lower temperatures in the post-liquefaction phase when a second dose of the alpha-amylase enzyme is added. In the conversion step, when the starch becomes dextrin and the pH decreases, the malted beta-amylase enzyme with malt acts in the same way as the commercial glucoamylase, which sometimes reduces the saccharification period. Since the quality of barley in the household malted barley may vary, it may be necessary to add a little bit of dosage to improve the efficiency of the malt. In addition, 5% of malt can be used if the enzyme activity is slow, but the total ratio of malt to lard should not exceed 20%.

5.7   Testing Procedures

Fermentation or anaerobic conversion of sugar to ethanol and yeast is a complex process that lasts from

24 to 72 hours. Since each raw material (feedstock) has a slightly different sugar or starch content, and often both, it is not always easy to determine what percentage of sugar is available after fermentation when the conversion is complete.

The amount of sugar that is fermented in the mash and more, how effectively they consume by the yeast, is crucial to determine the performance of the alcohol. As an ethanol manufacturer, it is to your advantages by seeking the best balance between content and economy in the raw material (feedstock), and also to ensure that as much material as possible becomes alcohol as much as possible. The only correct way to determine the content of the raw material is an analysis, which is not always practical for a small producer (although there are other ways that we see at the same time that are much cheaper and almost as effective).

However, if you start with a certain amount of raw material (feedstock) and a known amount of carbohydrates, you will know what to expect with respect to ethanol production performance and when the conversion is complete. If the conversion does not produce what you need, you can work on recognizing and solving problems.

5.8   Testing for Starch-to-Sugar Conversion

If you do not work with fruit or sugar stocks, it is likely that the raw materials (feedstock) you will use are a starch-based. The use of the term "carbs" in the food sense can lead to confusion, as many believe that carbohydrates are synonymous with starch. In fact, carbohydrates are a generic term for organic compounds composed of carbon, hydrogen, and oxygen, which also include starches, sugars, and celluloses. The conversion of starch into sugar takes place in two stages.

The first stage, liquefaction, uses alpha enzymes (alpha-amylase) and transforms a long chain of polysaccharides, a combination of 11 or more monosaccharide, or simple sugars held together by glycoside bonds to dextrin and polymers of intermediate glucose in the complexity between starch and maltose sugar. Few manufacturers can test at this stage, except for empirical observation: if the mash is thick and gelatin after cooking, a large amount of un-hydrolyzed starch may remain in the mixture, indicating that the initial conversion phase is incomplete.

If properly prepared, the mash will pass through the degree of gelatinization during which the stem cell walls decompose and the mixture thickens but remains liquid during the cooking process. This first step of starch conversion has to be completed, as the next step takes place under significantly different conditions.

The second step, saccharification, is the final transformation of dextrin into simple sugars or glucose with another set of glucoamylase or beta enzymes. Temperature and pH at this point are lower and must remain so that enzymes act. The progress of this conversion can be estimated using a simple iodine test. The starch responds with iodine turning into deep purple; the partial presence of starch is blue; a light red or yellow tint indicates no starch, indicating that most or all of the starch has been successfully hydrolyzed.

To prepare the iodine solution for testing, you need to buy a Lugol's solution called "Strong iodine solution" (USP) from your pharmacy or chemical supplier. Mix

1 ml of this solution with 7 ml of distilled water. You can use standard iodine in the pharmacy and dilute it with distilled water at a ratio of 1: 9. Both solutions have a short shelf life and should be used within 24 hours of mixing.

The test is carried out first by filtering a small amount of wort through a coffee filter to remove the solid particles to obtain virtually colorless liquid. Mix 20 ml of this filtered wort must with 30 ml of distilled water to get 50 ml of a small volume of liquid at a ratio of 2: 3. Add a solution of iodine in small amounts with an eyedropper, which means a change in wort color sample when introducing each drop.

If the color of the sample does not change with the addition of iodine, you have achieved a result. As already mentioned, the yellow color is the ideal conversion. The deeper, especially blue and purple, indicates that it has not yet transformed - and thus lost – starch in the mash.

As hard as it looks, it is always good to monitor the results in these tests to demonstrate consistency that is to keep a record of everything. Use the same amount of liquid and note the number of drops used as well as the color detected. This is especially useful when trying different raw materials (feedstock) and enzymatic effects to get the best results for production.

In addition, the iodine method can be used to test the efficiency and dose of enzymes, especially in unidentified raw materials (feedstock). Enzyme manufacturers optimize their formulations and dosages for a particular raw material (feedstock), and a small scale manufacturer may not work with common commodities of feedstock. Regardless of what you are using, you should try to determine as accurately as possible the carbohydrate content in your inventory using the available resource information. However, this is only an educated assumption.

If you make a sma5.9 ll sample for testing, then set about half a dozen of the test kit sets with different dosage enzymes, you can determine which works best. A sample should be kept under dose-controlled monitoring recommended by the manufacturer. Others can use relatively fewer enzymes (do not forget to store the doses). Keep all sample containers at the recommended temperature to facilitate the enzyme function and periodically rotate the contents to agitate.

After an hour, when the enzymes are active, cool the sample tanks and continue the iodine test, adding the droplet number of solution to each sample uniformly. You may find that more or fewer enzymes are needed to get the best results from a particular raw material (feedstock), as the manufacturer has stated. It is best to use the one that completes the conversion best.

5.9   Testing for pH Levels

The "PH" symbol is used to indicate an effective hydrogen ion concentration or simply the acidity or solubility of the solution. The scale 0-14 represents the available range, 7.0 neutral values. Each value below 7 indicates the acidity while values greater than 7 indicate bases or alkalinity. The PH scale is not linear - it is an inverse logarithmic representation, i.e. each pH unit differs 10 times from the next unit. Thus, the transition from 3 to 5 is 100 times greater than the transition from 3 to 4.

Since yeasts and enzymes are well-defined areas in living organisms where they can survive, pH is important. Some areas may be tight, and when microorganisms environment turn out to be ideal, their activity slows down which they won't work efficiently and even be destroyed. The enzyme manufacturer must provide you with a technical data sheet on the storage, handling, and use of the enzyme product. The PH value must be tested in every production and fermentation phase. The water used in these processes naturally affects the pH, as well as the type of raw material (feedstock), used and to some extent the material handling equipment.

In the case of water, it is worth noting that well-treated water and also community material are likely to contain chlorine and other additives such as fluorides, which can be detrimental to yeasts and enzymes. It would be prudent in a qualified lab to test the sample of your intended water to see if there is a potential problem.

Simple fecal coliform tests by the Health District are not suitable for your production. More complex analysis is required which includes numerous parameters such as sodium, cobalt and said chlorine. National Testing Laboratories or the State University with the Environmental and Health Program can provide you with a schedule of available testing levels.

Appendix B contains a list of national courier services. If you work in rural or open spaces, you probably have access to a well or spring water that can be contaminated with chemicals or fertilizer water, but not chlorine. Again, a reasonable price is a water test at reasonable prices. The simplest means of testing the pH is the litmus paper. Paper is sold in narrow strips and reacts with the solution to change the color.

By adjusting the hue to a color chart that came with the paper, you can quickly determine the pH value. Indicators are sold in the complete range of 1 to 14 or are divided into narrower areas such as 3.0 to 5.5, 3.4 to 4.8 or 4.5 to 7.5. The narrow-range papers are more accurate because it is easier to determine the exact value.

The use of litmus paper is pretty easy. Remove the tape from the container and immerse the other end in the sample solution. It immediately changes color and can be immediately compared to the color table. Make sure they have enough light to pinpoint colors. If you leave the paper in direct sunlight (or in a damp environment) the test strips will be lost (or darker) and the measurements will be discarded. Store the paper in a dry, uncovered environment. At a low cost, you can skip the paper and buy a pH meter. It costs hundreds of dollars or more, but even cheaper models are more accurate and offer fewer options than show papers. The meter has a wire that reads directly from the sample solution and indicates a value on an analog scale or a digital value. Some are calibrated with distilled water and others with calibration solutions. Each salt value of the meter is calibrated. Please note that the type of pH meter used for soil analysis in the garden is not suitable for testing mash solutions.

In determining the pH of the solution, it can be adjusted by adding lime (calcium oxide or agricultural calcium hydroxide) if it is too acidic or one of the various common acids (sulfur or phosphorus) if it is too alkaline. Muriatic or hydrochloric acid should not be used as they may be toxic to yeast. Both acids and bases must be handled with safety equipment and caution.

5.10 Testing for Sugar Concentration

Since sugar is the base from which ethanol is produced, this is essential for the process of fermentation. Whether it is a simple sugar (monosaccharide) or a disaccharide (combined with two monosaccharide units), it can be fermented. Feedstock that is predominantly starch usually does not require sugar measurement since your end result is to know if the starch is successfully converted and the iodine test can be used to determine it. However, in raw materials (feedstock) such as sugar cane, cheese, and remains of sugar beet or sugar beet molasses - you need to know the percentage of sugar used to calculate the ethanol yield.

The concentration of sugar can be easily tested with a saccharometer a hydrometer which is Brix scale calibrated. It measures the concentration of sugar in water based on the density of the liquid. The saccharometer is available in various ranges 9 to 21 percent, 18 to 35 percent, or 0 to 25 percent. The hydrometer clip floats in the solution at the scale embedded in the pane. The reading is performed at a point where the fluid level hits the shaft partially under water.

All solids in the solution drop to readings. It is, therefore, best to filter the liquid more than once through a paper filter to remove insoluble solids. Saccharometer is also calibrated at a specific temperature of -60 ° F; Readings above the calibrated temperature show less sugar than is actually available.

The Saccharometer is cheap but gives inaccurate for the reasons mentioned above. For accurate measurements within a fraction, percentage requires a refractometer, a device that measures the curvature of light through a sample of transparent sugars using prisms and lenses mounted on tubes (cost: about $ 20). The light wave is refracted and projected to the eyepiece; the Brix scale shows a separate line at the percentage point. The refractometer is also sensitive to temperature, but the higher state of ATC (Automatic Temperature Control) even compensates for the difference. If you really want to produce alcohol fuel, investing in this sophisticated tool will give you a better return by reducing your sugar waste. To put things in perspective, you can expect one thousand pounds of raw material at 18 percent or more of sugar to get 14 liters or more of alcohol.

5.11   Testing for Alcohol Content

Your final ethanol product must be tested to ensure its durability and strength, not only to satisfy your curiosity, but also to determine the quality of ethanol used as fuel. More importantly, the Federal Office of Taxes and Trade (ETB) for Federal and Tobacco require you to sign up for the full amount and proof of the fuel it produces. The proof strength value, which is just below the azeotropic levels of 190 to 192, is close to the finalization of the direct ethanol fuel use. To obtain such evidence beyond the expectation require expensive equipment and beyond the capabilities of a small scale producer. Generally, 160 to 185 samples are accepted as to fuel for the vehicle, at least they are lit in the combustion chamber, but water becomes a problem.

For 80% ethanol and 20% water (160 tested), fuel consumption is significantly reduced. The corrosion factor induced by electrolysis is of particular concern, particularly in the presence of zinc and aluminum, two common metals in engine parts. Problems with cold start and freezing winter crystals are also a problem in a less proof fuel. All of these problems will effectively disappear when the proof strength is 185 or so.

Like the saccharometer, the Proof & Tralles meter measures the specific gravity of the solution, in this case, alcohol and water. The test is calibrated on a scale of 0 to 200, and the Tralles scale represents 0 to 100. The test meter (hydrometer) is also calibrated at 60 ° F and gives false readings above temperatures or below this temperature. Larger liquid temperatures lead to higher test readings. Therefore, the "real percentage" correction map published by the TTB is used to calculate the actual proof at any temperature between 61 ° F and 100 ° F. This prevents each sample from cooling from the test taken. It is also possible to measure the proof strength with a refractometer, but the Brix scale number must be converted to a refractive index and then to real proof. For simplicity, a refractometer is available that is calibrated to use the refractive index scale instead of the Brix scale.

5.12   Starch Crop Fermenting Recipes

On a small scale, corn may not present such issues locally, especially if it's available as surplus, or grown and harvested less intensively. What follows is a conventional recipe for fermenting corn and other grains which we used successfully in our alcohol fuel program, using at the time the Biocon (US) enzymes Canalpha (alpha amylase) and Gasolase (glucoamylase), and the company's Special Distiller's Yeast (*S.cerevisiae*). (Biocon (US) Inc. has since become part of the firm that manufactures Novozymes; the Biocon (US) enzymes were packaged in 2.2-ounce measures and the yeast in 8.8-ounce packets.)

5.13   Conventional Corn Mash Recipe

MILLING AND PROCESSING

Chaff, or pieces of cob and stalk, should be removed from the grain before grinding. The milling can be done with a hammer mill screened to produce a coarse to medium grind (not a fine flour grind).

SLURRYING

Add 20 gallons of clean water per 1 bushel (56 pounds) of ground grain. Bring the temperature to 150°F. Begin agitation and add the grain slowly, to minimize clumping. Adjust the pH to the range of 5.5 and 7.0.

LIQUEFACTION

Add 1 packet of Canalpha for each 5 bushels of ground grain. The powdered Canalpha must be premixed to a paste with 100°F water, and then further diluted to a thin liquid before introducing it to the wort. Over a period of 30 minutes or more, raise the temperature to boiling and allow the wort to boil for 15 to 30 minutes.

POST-LIQUEFACTION

Cool the mash to 160°F by adding 5 to 10 gallons of cool water per 1 bushel of ground grain. (The manufacturer recommends 25 to 30 gallons of water per 1 bushel of ground grain as the final ratio.) Add 2 packets of Canalpha (again, first made into a paste) per 5 bushels of ground grain and hold for 30 minutes.

HYDROLYSIS FLOW CHART

Conventional Hydrolysis

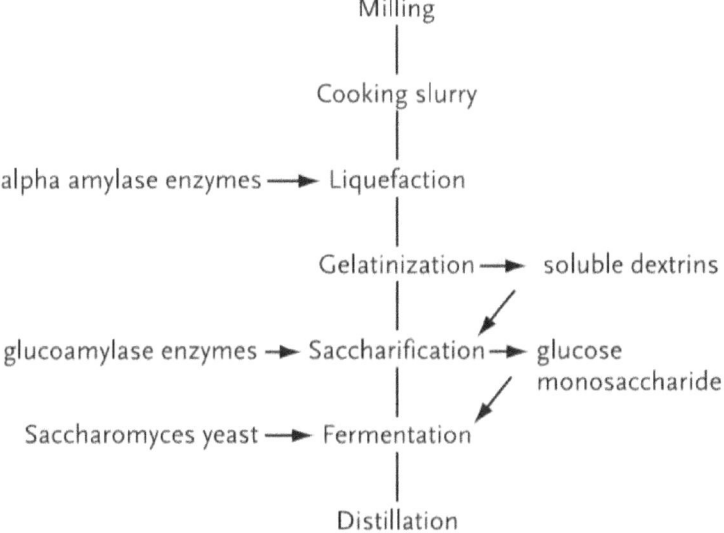

Milling
|
Cooking slurry
|
alpha amylase enzymes ⟶ Liquefaction
|
Gelatinization ⟶ soluble dextrins
|
glucoamylase enzymes ⟶ Saccharification ⟶ glucose
| monosaccharide
Saccharomyces yeast ⟶ Fermentation
|
Distillation

Photo from: R. F. (2009). *Alcohol Fuel A Guide to Making and Using Ethanol as a Renewable Fuel.*

Conversion

Cool the wort to 90°F using cooling coils and add 1 packet of Gasolase and 1 packet of Special Distiller's Yeast per 5 bushels of ground grain. The powdered Gasolase and the Special Distiller's Yeast must be premixed to a paste with 100°F water, and then further diluted to a thin liquid before introducing it to the wort. After the wort is well mixed, stop agitationand transfer the mash to the fermentation tank, if used.

FERMENTATION

The mash will ferment over a period of two to three days, and will then be ready for distillation

5.14 Simultaneous Saccharification and Fermentation "No Cooking" Corn Mash Recipe

The advancement of enzymatic technology has resulted in granular starch hydrolysis enzymes such as those developed by Genencor International. The company's Stargen ™ 001 enzyme contains alpha-amylase and glucoamylase, which hydrolyze the granular starches of corn and other starchy raw materials without undergoing the gelatinization or liquefaction steps although wheat, barley, and rye may be one step of pretreatment of the enzyme required to reduce masculinity). The advantages include the rationalization of the fermentation process and lower energy consumption combined with high ethanol yields. In the next section, you will find a typical recipe of Stargen ™ 001 for the use of corn.

MILLING AND PROCESSING

Mill and clean the kernels through a 30 mm (0.023 inches) screen. The Stargen ™ 001 contains detailed information on the effects of different mesh sizes. A finer particle size leads to a higher alcohol yield but makes it difficult to recover. Small scale producer may need to work on an 1125-inch (1/8) screen.

SLURRYING/HYDRATION

Add 40 liters of water from 68 ° F to 104 ° F per 100 pounds of ground grain. Mix well

SACCHARIFICATION AND FERMENTATION

Add enzyme while retaining agitation. Use at least 2.2 pounds of Stargen ™ 001 2,200 pounds of ground grain (1.6 ounces of 100 pounds of corn). In some cases, up to 5.5 pounds of the dose of 2200 pounds of grain (4 ounces per 100 pounds) may be required. Adjust the pH of the mash to 4.0 - 4.5.

Add yeast or own yeast with higher alcohol tolerance (> 15%) within 15 minutes of pH adjustment. Continue mixing and maintain the mash at 90 ° temperature. The fermentation is completed within 90 hours. (Temperature requirements may vary from

86 ° F to 104 ° F and fermentation times may vary from 24 to 90 hours depending on the conditions.)

5.15 Potato Stock Mash Recipe

Raw potatoes have some properties that make them the ideal raw material. They have high starch content (almost 80%), high water content and are well supplied. Their starch granules are bulky and easily hydrolyzed, making it easier to convert sugar. They also have a low content of solid matter, which reduces the need for renewal of the spent mash.

The main disadvantage is the cost that can be compensated by using agricultural or by processing culls or increasing their activity. Another difficulty in treatment is that starch particles must be physically dissolved (heat alone won't be sufficient) and cellulose fibers must absorb liquid and must be compressed or extracted to increase the potential.

The following recipe was provided by Clarence Goosen, a former fellow, from is extensive ethanol research during and after the alcohol intake program. Taka-Therm® and Diazyme® enzymes are refer to products produce from Miles Laboratories, Inc., 1995 by Bayer AG. They are replaceable with any suitable alpha amylase and glucoamylase enzyme.

GRINDING AND PROCESSING

Grind the potatoes to its smallest particle possible. If you use the pump as part of agitation cooker, change the size up to 1/8 inch to avoid overloading the rotor. When the raw material is gelatinized, the centrifugal pump rotor breaks the particles into smaller pieces until they are broken effectively.

SLURRYING

The share of potatoes in water is about 80%, so a little extra fluid is needed to promote the cooker flow. Add 1 gallon of pure water to all 80 kilograms of raw potato. This leads to a 13% sugar solution in the last mash.

Adjust the wort to pH level 6.0 to 6.5 with a 10% lime slurry solution.

LIQUEFACTION

0.5 ounces of Taka-Thermia (alpha-amylase) is added to each 100 pounds of potato in wort. Stir continuously and raise the temperature to 145 ° F until the starch begins to gelatinize. Gradually add heat to keep the wort liquid. If the heat is added too quickly, the starch thickens and becomes too viscous for mixing. If this happens, lower the heat and allow the enzymes to continue to gelatinize the starch. Continue heating until the temperature is 194 ° F and hold for 30-60 minutes.

Test for adequate hydration by performing an iodine (starch) test in a 1 pounce sample of wort. Introduce a drop of iodine, mix and dilute with 10 ounces of clean water. The presence of any dark blue or purple particles in the solution means that the starch is not completely liquefied and the starch fragments are not sufficiently disintegrated. If so, continue the cook until there are no more black particles in the test. Remove the heat source and add another 0.5 ounces of Taka-Therm-alpha-amylase enzyme to 100 pounds of potatoes for post liquefaction.

CONVERSION

When liquefaction is complete, reduce the temperature of the wort to 140 ° F by cooling with a heat exchanger or internal cooling coils. At this point, do not add cold water. Adjust the pH to 4.2 with sulfuric acid, and then add 0.75 ounce of diazyme (glucoamylase) to 100 pounds of potatoes. Keep the temperature at 140 ° F for one to two hours.

FERMENTATION

Cool the wort to 86 ° F and disconnect the liquids from any solid. Add yeast and keep at 86°F. Within 48 to 72 hours fermentation will be complete.

5.16 Basic Sugar Technique

In the earlier Chapter, I explained that the structure of some non-complex sugar structures in some crops that are easy to ferment without going through hydrolysis process. These so-called Sugar crops, including sugar beet, sugar cane, fruit and sweet sorghum, can be excellent raw materials at a low price if they are processed as soon as possible after harvest to avoid damage to the feedstock.

As with starch, sugar technique can be divided into the following stages: (1) extraction, (2) cooking, (3) cooling, and (4) adjusting the sugar, pH and nutrient content.

But cooking or sterilizing the sea can be expensive in large quantities. An alternative solution is to use a low pH associated with high doses of yeast to remove unwanted bacteria that are normally controlled by sterilization.

You can consider several options when you think of sugar as a raw material (feedstock). Because sugar-based raw materials (feedstock) are generally not good for a long time, they are distilled as soon as they are produced and are usually produced seasonally, near the harvest or during harvest.

Traditionally, sugar crop are "juiced" by crushing or pressing them into the machine. The sugar cane typically crumbles in a drum crusher or roller mill that squeezes and collects juice and sends the stock out of the transport or drying tubes. The juice extract has high sugar content because the liquid is already concentrated and does not require additional water. A disadvantage of this technique is that up to a quarter of the sugar in the plant remains in the press - however, the sugary residue is also important as a fertilizer or food, especially in fruit crops. Crushers are produce in all sizes.

They are also available as agricultural machines or as salvage. A twisted press is just another, more sophisticated version of the crusher. However, these presses are generally limited to industrial use and can be very expensive. A simple arbor press can be the best option for a small scale producer. Hydraulic types are 25-ton cylinders mounted on a vertical rigid frame and the piston downward. It is designed to detect (or connect) bearings, needles, and rollers on machines, but it can be easily adapted to compress pulp or crop by welding a wider end to the piston and attaching the sturdy receiver underneath. The container or reservoir is provided with an outlet or a closed bottom for juicing. The simplest hydraulic presses work by hand (manually). They have larger electric pumps and hydraulic valves as well as two-way cylinders so that both extension and retraction are powered.

There is also a juice process, called diffusion, in which the pulp or thin sections of crop are heated to 190 ° F and soaked in a bowl to remove the sugar from the substrate. The enriched solution is then passed through several degrees of heating and steeping to gradually extract the sugar from the fresh substrate into a series of containers. Once the original sugar intake is exhausted, it is removed for use as a bagasse or livestock feed, and fresh feedstock is added. On this rotating substrate, a new substrate was added until supply is exhausted. A completely different approach, which can also be easily done on a low scale, is the crushing of the raw material (feedstock) in a hammer mill or shredder.

Although the pulped material is more problematic in handling, it is used everywhere and the spent mash can still be used to feed livestock. Since the water has to be added to the pulp to be fluid enough to pump or mix the original sugar concentrated is somewhat lower than that of the straight juice, but it is still quite acceptable. Such pulped wort can benefit from a higher dose of yeast and a good mix because the yeast is not as mobile in a thick solution. Medium-sized shredders are common in landscaping and gardening operation, and are not as expensive as a hammer mill.

Step 2: Cooking...or Not

Cooking is the best option to determine the economy of your heat source. In fact, one of the benefits of using sugar as raw materials is that it can eliminate the hot cooking phase, as I just mentioned, to save fuel costs. But, let suppose you use wood or biomass as a heat and decide to cook it for successful sterilization. Even a small shredder can be useful to produce feedstock preparation. The shredding teeth, which are seen in detail grind and even separate hard raw material (feedstock).

Temperatures should be maintained at a temperature above 200 ° F for one hour during agitation. High temperatures destroy harmful bacteria and also help to crush the pulpy material if you have shredded the raw material instead of pressing it. It is possible to avoid cooking at high temperatures simply by heating the wort to 90 ° F and lowering the pH if necessary to 4.0 or 4.5 by adding lactic acid or sulfuric acid. Lactic acid supports the growth of yeast and still inhibits the growth of harmful bacteria.

After completing the pH and other adjustments listed below, you can bring in a large inoculum of dry yeast (more than one pound per gallon) thus, competes effectively with food bacteria. A higher dose ensures at least in theory that yeast will prevails in this competition. It is important that temperatures do not exceed the yeast tolerance level at 90 ° F. The ideal area for yeast is between 78 ° F and 86 ° F, but the ambient temperature also affects this situation. In an environment lower than room temperature, wort of

90 ° F is preferred because of the loss of atmospheric heat. In warm weather, the wort must cool down to 65 ° to make the best environment. In any case, yeast metabolism generates heat to increase the temperature.

Step 3: Cooling

If you've cooked your wort, you have to cool before you add yeast. Again, the purpose is to get it in a confortable range for the yeast from 85 ° F to 90 ° F. Do not add cold water to cool the mix, since it will definitely dilute the wort and almost certainly change the pH level. It is better to use a tubular cooler or heat exchanger in a container that circulates cold water and reduces the temperature without additional water.

Step 4a: Sugar Concentration

The sugar concentration of the wort should be set at 10-20%. Although it seems that higher sugar concentration will eventually produce more alcohol, this statement is only worth to a point. In fact, the alcohol content in the wort exceeds the yeast tolerance, actually modifies the structure, limits its activity and finally prevents all available sugars from becoming ethanol.

The low sugar concentration, in turn, allows an uneconomical use of the fermentation quantities and prolongs the duration of the distillation, which cost on the long term. One of the benefits of keeping the sugar concentration at the bottom of the scale (12 to 14% range) is to ensure that most sugars will be fermented. It also protects against overheating due to the energetic metabolism of yeast in a very high sugar environment.

In the case of crushed raw materials (feedstock), it is possible to achieve higher sugar concentration by limiting the amount of water added to the preparing the wort. (Juices generally have higher content than crushed raw materials (feedstock) because water has not been added during preparation). An again, since the yeast can only work with available sugar, it is safe to aim for a higher concentration of sugar with the raw material (pulp feedstock), because initially a large amount of sugar is still linked to the pulpy shreds. As the yeast exhaust sugar, more sugar is released over time, but not all at once, so that the yeast is only overwhelm with saccharides.

Sugar levels up to 22% are generally safe with raw materials (pulpy feedstocks). The exact sugar concentration can be measured with saccharometer or better still a refractometer (see "Sugar concentration test"). Any insoluble solid in the wort can affect the sugar readings, so it is best to test the unbleached paper filter before performing the test.

Step 4b: Adjusting pH Levels

The most common acid used to lower the pH in a solution that is too alkaline is sulfuric acid and phosphoric, also with lactic acid also an option. If the wort is too acidic, an agricultural lime or caustic lime (calcium oxide) should be introduced. The aim is to achieve a pH between 4.0 and 4.5. You can calculate the amount of acid or an additional base you should use, withdraw a small sample of your wort, like three gallons from 300 gallons. Carry out a litmus test or use a pH meter to read, carefully measure the amount of acid or lime is required (in milliliters of liquid or dry gram) to correct the solution to the appropriate level. Then you can extrapolate that amount to match total quantity of the wort. If it took 60 grams of lime to correct three gallons of wort, it will takes 5.940 grams (5.94 kg or 13 pounds) to correct the remaining 297 gallons (60 x 297 = 17 820 divided by 3 = 5.940). The sample protects you from "speculation" with attempts and mistakes throughout the set.

Step 4c: Adding Nutrients

Many sugar proteins do not contain all the minerals needed for yeast metabolism. Yeast products, in the form of ammonium sulfate or phosphate are usually added to wort to achieve complete fermentation. These ammonium salts contain phosphorus and nitrogen, which are needed by the yeast and two or three pounds per 500 gallons of wort are what is required. Calcium sulfates (gypsum) fed at the same speed as the fertilizer salts, produces the required calcium. Other nutritional supplements include yeast foods, but a more advantageous alternative is simply to add dry beer yeast to the mixture of five pound can; it provides nutrients outside of the mineral content.

Elsewhere in this chapter I have mentioned using up to one-third spent mash from a previous run to compensate for the fresh batch. This creates a yeastsource for fresh wort, but there is also a risk that the entire batch will be contaminated with bacteria if the substance used is not used immediately after distillation. When the time is considerable, the bacteria will have the perfect opportunity to multiply.

5.17   Pitching the Yeast

Once the wort has cooled and all adjustments made, the yeast is introduced to start the fermentation. Unlike the massive yeast dose described in the "uncooked" version of Step 2, sterilized wort require a less aggressive dose because they don't have to compete with bacteria for food. A typical method for "pitching" yeast (described in detail in the section "Preparation of yeast starter") is to remove about 5% of the sterilized wort and mix it at a yeast level of about 1/2 pound dry yeast per 100 gallons of fluid, agitate it and allow to create a fast-growing healthy colony before taking it back to your main wort.

The addition of free oxygen or injection into the wort is an important factor in the development of rapid early fermentation. Oxygen promotes growth and proliferation and can easily pass through a micro-bubbler tube consisting of a portion of a pre-drilled plastic tube attached to the oxygen tube and embedded in the fermentation vessel. Oxygen compressed by the welding system is ideal for this. The use of pressurized air is a cheaper option, but the air does not contain much oxygen.

Injection should only be done in the first 15 minutes so the yeast can then support up to 24 hours to absorb and proliferate oxygen. As a result, the yeast colony goes into the anaerobic phase and produces a significant amount of carbon dioxide (and alcohol) and CO_2 will bubble out vigorously in the fermentation lock, preventing air from being drawn into the enclosed vat. When the fermentation is complete after 24 or 36 hours or more, the bubbling will stopped and the mash is immediately distilled to prevent bacterial growth, thereby converting the mixture to acetic acid and rendering it unusable for ethanol.

5.18 Calculating Alcohol Yield

The ethanol yield by weight of the raw material (feedstock) is an important factor, not only to determine if the crop or substrate you are considering is worth it, but also to determine if your fermentationand distillation practice is right for you. The simplest way is to calculate the percentage carbohydrate content in the substrate, as explained in "Analysis the feedstock." Although this is only an estimate, it is pretty close and handy as it works with both starch and sugar crops.

Warning: The calculation is based on the complete conversion. Realistically, you cannot expect an almost complete conversion, as you probably cannot convert all starch or cannot absorb all sugars. Elsewhere, I used an 85% number to achieve my goal, so you can stick to that.

Crop	Percent Water	Percent Protein	Percent Carbohydrate	Crop	Percent Water	Percent Protein	Percent Carbohydrate

Water, Protein, and Carbohydrate Content of Selected Food and Farm Products

Crop	Percent Water	Percent Protein	Percent Carbohydrate	Crop	Percent Water	Percent Protein	Percent Carbohydrate
Apples, raw	84.4	0.2	14.5	Okra	88.9	2.4	7.6
Apricots, raw	85.3	1.0	12.8	Onions (dry)	89.1	1.5	8.7
Artichokes, French	85.5	2.9	10.6	Oranges	86.0	1.0	12.2
Artichokes, Jer.	79.8	2.3	10.6	Parsnips	79.1	1.7	17.5
Asparagus, raw	91.7	2.5	5.0	Peaches	89.1	0.6	9.7
Beans, lima (dry)	10.3	20.4	64.0	Peanuts	5.6	26.0	18.6
Beans, white	10.9	22.3	61.3	Pears	83.2	0.7	15.3
Beans, red	10.4	22.5	61.9	Peas, edible pod	83.3	3.4	12.0
Beans, pinto	8.3	22.9	63.7	Peas, split	9.3	1.0	62.7
Beets, red	87.3	1.6	9.9	Peppers, hot chili	74.3	3.7	18.1
Beet greens	90.9	2.2	4.6	Peppers, sweet	93.4	1.2	4.8
Blackberries	84.5	1.2	12.9	Persimmons	78.6	0.7	19.7
Blueberries	83.2	0.7	15.3	Plums, Damson	81.1	0.5	17.8
Boysenberries	86.8	1.2	11.4	Poke shoots	91.6	2.6	3.1
Broccoli	89.1	3.6	5.9	Popcorn	9.8	11.9	72.1
Brussels sprouts	85.2	4.9	8.3	Potatoes, raw	79.8	2.1	17.1
Buckwheat	1.0	11.7	72.9	Pumpkins	91.6	1.0	6.5
Cabbage	92.4	1.3	5.4	Quinces	83.8	0.4	15.3
Carrots	8.2	1.1	9.7	Radishes	94.5	1.0	3.6
Cauliflower	91.0	2.7	5.2	Raspberries	84.2	1.2	13.6
Celery	94.1	0.9	3.9	Rhubarb	94.8	0.6	3.7
Cherries, sour	83.7	1.2	14.3	Rice, brown	12.0	7.5	77.4
Cherries, sweet	80.4	1.3	17.4	Rice, white	12.0	6.7	80.4
Collards	85.3	4.8	7.5	Rutabagas	87.0	1.1	11.0
Corn, field	13.8	8.9	72.2	Rye	11.0	12.1	73.4
Corn, sweet	72.7	3.5	22.1	Salsify	77.6	2.9	18.0
Cowpeas	10.5	22.8	61.7	Soybeans (dry)	10.0	34.1	33.5
Cowpeas (undried)	66.8	9.0	21.8	Spinach	90.7	3.2	4.3
Crabapples	81.1	0.4	17.8	Squash, summer	94.0	1.1	4.2
Cranberries	87.9	0.4	10.8	Squash, winter	85.1	1.4	12.4
Cucumbers	95.1	0.9	3.4	Strawberries	89.9	0.7	8.4
Dandelion greens	85.6	2.7	9.2	Sweet potatoes	70.6	1.7	26.3
Dates	22.5	2.2	72.9	Tomatoes	93.5	1.1	4.7
Dock, sheep sorrel	90.9	2.1	5.6	Turnips	91.5	1.0	6.6
Figs	77.5	1.2	20.3	Turnip greens	90.3	3.0	5.0
Garlic cloves	61.3	6.2	30.8	Watermelons	92.6	0.5	6.4
Grapefruit pulp	88.4	0.5	10.6	Wheat, HRS	13.0	14.0	69.1
Grapes, American	81.6	1.3	15.7	Wheat, HRW	12.5	12.3	71.7
Lamb's quarters	84.3	4.2	7.3	Wheat, SRW	14.0	10.2	72.1
Lemons, whole	87.4	1.2	10.7	Wheat, white	11.5	9.4	75.4
Lentils	11.1	24.7	60.1	Wheat, durum	13.0	12.7	70.1
Milk, cow	87.4	3.5	4.9	Whey	93.1	0.9	5.1
Milk, goat	87.5	3.2	4.6	Yams	73.5	2.1	23.2
Millet	11.8	9.9	72.9	Note: From the Handbook of the Nutritional Contents of Foods, USDA.			
Muskmelons	91.2	0.7	7.5	Values may differ slightly from online National Nutrient Database values			
Mustard greens	89.5	3.0	5.6	due to continuing research.			

Photo from: R. F. (2009). *Alcohol Fuel A Guide to Making and Using Ethanol as a Renewable Fuel.*

Firstly, determine the carbohydrate content in Table above or visit the USDA website in the "Analyzing" section. Then, take the figure as a whole number and divide it by 13.5. The result is the alcohol quantity that can be expected per 100 pounds of substrate. You can turn it into a ton by simply multiplying 20. I will use a corn as an example. The carbohydrate content of Dent corn is 72%, 72 divided by 13.5 = 5.33 gallons per 100 pounds. Simply, One ton equals 20 times 5.33 or 106.6 gallons. The correction of the actual or real life scenario is you will multiply by 0.85, giving 90.6 gallons, which is quite extensive in the corn ethanol range.

R. F. (2009). *Alcohol Fuel A Guide to Making and Using Ethanol as a Renewable Fuel.* Canada, New York: New Society. **Retrieve** from: www.newsociety.com

6 DISTILLATION PROCESS

6.0   How Distillation Works

The distillation process is simple, but its details are complex. Its purpose is to separate the alcohol from a mixture of alcohol and water containing the fermented mash liquid. In general, liquid mash or "beer" contains 10 or 12 percent alcohol. Therefore, almost 90% of the volume of water must be removed to obtain ethanol containing fuel grade ethanol. In an open atmosphere, water boils at 212 ° F and ethanol boils at 173 ° F. Alcohol has a higher vapor pressure than water, which means it uses less energy to convert it from liquid to gas. When mixing alcohol and water, the boiling temperature varies and is between the boiling points of the individual components.

The higher the ethanol concentration of the mixtures the lower the boiling point. On the contrary, the less ethanol is present in the mixture, the higher the boiling point, since the distillation process produces a change of phase from liquid to boiling (vapor). The more ethanol is extracted from the mixture, the higher the temperature to maintain steam. The following table shows the temperatures required to make a mixture of ethanol and water in mixtures in various percentages.

Boiling Points for Alcohol/Water Mixture at 1 Bar Pressure		
Percent Ethanol	Percent Water	Boiling Point, Degrees F
0	100	212.0
5	95	203.3
10	90	199.4
15	85	196.3
20	80	194.0
25	75	192.2
30	70	190.4
35	65	188.9
40	60	187.7
45	55	186.4
50	50	185.0
55	45	183.4
60	40	182.3
65	35	180.8
70	30	179.4
75	25	177.8
80	20	176.3
85	15	175.4
90	10	174.2
95	5	172.7

Note: One Bar pressure is equivalent to one Atmosphere or 14.69 psi, standard pressure at sea level.

Photo from: R. F. (2009). *Alcohol Fuel A Guide to Making and Using Ethanol as a Renewable Fuel*. Canada, New York: New Society.

Let's look at a simple example of how distillation works. If we cook a mixture of ethanol and water in the container, it will produce more ethanol vapor than water vapor. If we take this vapor and the condensate it, the ethanol content in the liquid will be higher than that of the original mixture and the original mixture also contains less ethanol. Now, if we heat the condensed liquid, intercept its vapors and condense to liquid, the concentration of ethanol in the condensate will even be higher.

This method, called rectification, can be repeated until most of the ethanol is removed. In fact, this is how the first stills have increased the proof strength of the final product. But things are not that simple. The azeotropic state develops when the mixture reaches about 96% ethanol or the 192 proof. At this concentration, the ratio of ethanol molecules to water molecules reaches equilibrium, and methods other than distillation must be used to further enrich the condensate. These include the use of a dewatering device known as a molecular sieve that uses aluminum-silicon to separate water and carbon dioxide from ethanol, or a process that involves the addition of benzene to produce a ternary or triple azeotrope to reduce the boiling point to that of ethanol which is then separately extracted.

Although both methods are feasible on a large scale production, benzene is no longer used to dehydrate alcohol because it is a dangerous carcinogen. And the cost of operating a small-scale industrial molecular sieve since 99.9% pure alcohol is not needed to use motors or heating devices. The only reason alcohol fuel is distill to that grade is that, to be mixed well with gasoline when producing gasohol or E-85, 85/15% ethanol and unleaded regular. Anhydrous alcohol should be stored in closed containers or in containers with special ventilation openings.There are other cheaper methods to dry ethanol using limestone (calcium oxide) or rock salt (sodium chloride or calcium salts).

6.1   The Art and Science of Rectification

Old time stills, the redistillation process took place in separate tanks, but with modern equipment, everything is done in a fractionating column, which was a large pipe, sometimes called a tower. The column mainly allows you to stack a series of mini-still to be stacked with each other. The rectification is made in layers and each layer is determined by a ventilated plate. The first still were too complex to build. The plates were punched with tubes that extended a short distance over the surface of each plate. The tubes were covered with vented caps that allowed the steam to flow. Each plate also contained a lower angle, a longer tube suspended from the plate above it and ended in a well-built base plate on the surface of the plate.

Finally, manufacturers have developed simpler designs using perforated plates without detailed fabrication. These are chambers filled with pal l-ring or saddle packing shaped that allow them to reach a maximum surface while still allowing up to 90% of free space. These differential columns stills allow the transition between phases to occur in the package, at intervals where the vapors rise through the column. Packaged differential columns are the simplest to build, but because of the packaging, it cannot deal with solid in the mash.

Depending on the design of the still used, a number of plates or steps are needed to obtain a very durable ethanol product, which we can define as 190 proof or

95% ethanol test. In theory, a 10% mash mix should go through 25 steps to achieve that proof strength. But due to normal efficiency losses, sometimes it takes 44 or more plates to really get the job done. Some steps can be taken to increase the test of the ethanol produced in the distillation. The simplest is that the process lasts longer.

The nature of the distillation is that it takes about the same time to obtain a very durable product since the final 25% of the batch run as it does to distill the initial 75%. (A run is the period of time it takes to finish one distillation). The content of alcohol in mash is so low in the final stages about 2%, it takes extra time (and can be set up well) to achieve the same ethanol out of it as possible. At one point, energy consumption replaces the benefits and perhaps it's better to let the remainder pass through the column as "low wine" at 30 or 40 proof, separately collected and added separately to the next set of distillation when you are ready to make a run.

For reasons of efficiency, other methods have been developed that produce a high degree of purity without investing so much time and energy. One is to remove heat from the column using a heat exchanger. These are simply water-cooled coils that are strategically placed on the temperature monitor column at the critical points of the distillation process.

The second method is called reflux, which returns part of condensed ethanol back to the top of the column to increase conversion. Both methods can be used, or individually, depending on the design of the stills. Controls that are being made to these improvements are detrimental to costs and complications occurring in the production of ethanol, the most advanced models are needed to keep costs per gallon in range conventional fuels.

6.2   Continuous Distillation

To simplify the issues we have discussed so far, it is a batch distillation or a single-run process in which a predetermined amount of mash is prepared, heated and sent to a distillation column to produce the largest amount of ethanol from the system. When the maximum percentage of ethanol is extracted from the mash, the system is cleaned and the process is repeated.

The continuous distillation of the feed automates the process and facilitates the flow without stopping. Continuous still is more expensive to build than a batch still, partly due to the construction of the pillars, and partly due to the necessary sensors and controls required to make them work. The continuous distillation uses the multi-plate fractionation column described above, but feeds the fermented beer in its column at midpoint. Unlike the typical batch process, where the vapor enters the column, the continuous feed means liquid mash, sand of all the particles (therefore, compact columns rarely use continuous feed design unless the mash is pre filtered before the injection).

To resolve the clogging issue while maintaining a low costs, some models use perforated plates in the part of the column (section of the stripper) and in the package (rectifying part) of the section mentioned above, which only receives steam. The spent mash is maintained at a constant rate and which the still is kept at equilibrium.

Equilibrium is controlled by the amount of heat entering the boiler or the lower part of the stripper column. With the still at equilibrium, each plate is kept at its own boiling point and proof, which allows extraction of fusel oil and other contaminants in places where they stop evaporating in the column. This is an additional benefit for commercial operation that requires pure alcohol for medication or cosmetics. It is not difficult to overheat the pillar, especially with certain types of still design, or if the heat source is not consistent, such as in a wood boiler still. The effect is the same as when removing a panel or a phase, because once the column is above the optimum temperature, fractionation and equilibrium cannot occur. Therefore, the use of a heat exchanger to change the temperature inside the column is necessary to allow a minimum level of control.

Continuous feed columns are usually higher than the batch feed columns to adapt to the constant flow of material and achieve the necessary control for efficient distillation.

The vertical difference is such that continuous feed columns are often divided into two columns: the stripper and the rectifier, although they are treated as one. Vapor coming off the stripper column are channeled to the bottom of the rectifier so that they can increase the distillation, also the reflux of the fluid descending from the rectifier is pumped to the top of the stripper so that it can fall into the boiler to mixed up and vaporize with incoming mash.

One of the main advantages of continuous distillation is that a small amount of heat energy is consumed if properly recycled. When the process is complete, hot water from the re-boiler can be used to warm the water for mash cooker or preheat the fermented mash in the heat exchanger before going into the column. As expected, additional pumps and water supply systems are required which increases the cost of the system.

In general, small operations are better because of the lower cost and ease of handling operation by a batch-feed system. Here, "small scale" can be loosely defined, as in any 2-inch diameter micro-column that still generates less than one gallon of alcohol per hour in a 12-inch column at a output between 25 to 35 gallons per hour , (Doubling the column diameter increases the stills capacity by a factor of four). If you are targeting a volume larger than that, it is probably better to switch to the continuous-feed system and place it on monitors and controls, rather than spending money on larger diameter columns and additional costs to material.

6.3   Azeotropic Distillation

I have already mentioned that there is a point of diminishing returns where straight distillation of ethanol is concerned. At about 192 proof, the mixture of alcohol and water forms a strong bond that is separated only by heat and the mixture becomes what is called an azeotrope. I want to be clear on this: ethanol is ideal for most people interested in small-scale alcohol at that percentage. This alcohol is ideal for vehicles, ovens and gasoline equipment such as mowers, tractors, pumps and generators. The small amount of water in the fuel is usually not a problem for most people in most cases. Brazilian Flex fuel vehicles I refer to in chapter 1, have been running on hydrous alcohol fuel over half a decade.

In North America, Flex Fuel cars and trucks are working perfectly with hydrous alcohol. There are probably only two situations in which about 4% or so of the water in the fuel can cause problems. First, if someone lived in a climate where winter temperatures had fallen to -15 or -20 ° F, and ethanol-gasolineblend to power an unmodified engine to run on alcohol. The extreme cold would dissociate the fuel mixture into layers and the alcohol would soon pass through a fuel system that is not optimized for it. At an appropriate temperature, provided at 190 proofs gasoline will blend with hydrous alcohol without causing this problem

Column Diameter and Output Capacity		
Column Diameter, Inches	Capacity, Gal./Hour	Area, Sq./In
2	.75	3.15
3	1.5	7.07
4	2.6	12.56
5	4.2	19.64
6	6.0	28.27
7	8.2	38.49
8	10.7	50.27
9	13.5	63.62
10	16.7	78.54
11	20.2	95.03
12	25.0	113.10
Note: Alcohol yields at 190 proof.		

Photo from: R. F. (2009). *Alcohol Fuel A Guide to Making and Using Ethanol as a Renewable Fuel.* Canada, New York: New Society.

The only other reason why a person wants to produce anhydrous (free of water) fuel is to sell it as an anhydrous product. While this is certainly possible, it can be energy intensive, even with the help of convenient, complex and cheap techniques. However, since 99% pure ethanol is highly hygroscopic, it easily absorbs moisture from the air until it reaches a well-known azeotropic point. Anhydrous alcohol must be stored in closed, condensed-free containers with a special vacuum air, which will react only to significant pressure changes.

6.4 Drying Ethanol

One of the conventional methods for drying ethanol is the basic method of irrigation or transport through an absorbent material such as calcium oxide. Adsorption is surface retention of liquids, as opposed to absorption, where it is completely absorbed in most other substances. Calcium oxide or caustic lime (CaO) absorbs water, but not alcohol. Hydrogenated ethanol immersed in a reservoir containing limestone "slakes" the lime creating calcium hydroxide (Ca (OH) 2) filled with water. Calcium hydroxide is deposited on the bottom of the container, while almost pure alcohol is goes to the top. To remove 1 gallon of water from alcohol it takes about 35-40 pounds of lime therefore, it is important to determine the proof of your hydrous product and properly size the container: 100 gallons of 192 proofs ethanol contain four gallon of water and there must be room for fuel and the lime.

After soaking for 24 hours, the liquid can be extracted, distilled at 173 ° F and condensed. The substance remaining in the bottom of the distiller vat may be dried and regenerated to calcium oxide or redistilled liquid to obtain residual alcohol in the flakes. When the temperature value approaches the boiling point of the water, the alcohol content is drained and the liquid can be added to the next distillation run.

The remaining drying or re-distillation of the residual material for alcohol recovery is energy-intensive. The calcium hydroxide drying process requires temperatures in the range of 350 ° F, which cannot be performed on an open flame source. Dry heat delivered through the heat exchanger or flammable heated gases such as carbon dioxide should be used. Regenerated lime should not be exposed to open air or it will absorbed by the ambient moisture.

Sometimes it is better to use resources to leave the lime in its slake form rather than reuse it.

The second drying method is not as effective but can be acceptable on a small-batch scale. The hydrated alcohol can be run through a reservoir of calcium salt or rock salt which extracts water when it forms a desiccant. A simple gravity feed system is sufficient and the hydrated liquid can be collected through a port at the bottom of the container (it may need to be filtered to remove residue). The salt can be dried in hot air for reuse.

6.5 Choosing a Heat Source

Living in a rural environment, i will suggest you make use of wood as a fuel. Although natural gas, propane, heating oil or electricity is readily available, rural residents are attracted to at least some of their needs. Wood is renewable, it burns clean when properly burned with air and contains a lot of energy per weight.

Less than 1½ white oak cord can produces the same energy as 11.5 gallons of fuel oil. If you are a small scale producer and you needs a reliable and source of heat to power your equipment I will consider you make use of wood.

Any deficiency that diminishes the potential of wood for domestic heating, storage, inconvenience, dry heat is irrelevant if you use outside space or work building. Wood is probably not an option for an urban distillery or where it must be transported remotely.

Photo from: R. F. (2009). *Alcohol Fuel A Guide to Making and Using Ethanol as a Renewable Fuel.* Canada, New York: New Society.

However, in many parts of the country, wood has become a "value-added" commodity that holds a cost only once it's been processed? In raw form, on the stump or as logs, it can still be relatively inexpensive. Exurban and rural communities that are experiencing a surge in development often cannot dispose of logs unsuitable for milling (which is a large percentage of the total) without paying a tipping fee at the landfill. Public works and highway expansion projects that come with growth and development are also a source of cheap or even free wood.

The key to applying wood heat properly is in designing the firebox and cooking vat or mash pot in such a way as to expose as much of the vessel's surface area to the heat source as possible. This encourages the transfer of heat to the liquid within, though much of it will still escape out the flue pipe. Building an enclosure around the fire — a firebox — is critical to thermal efficiency to prevent the wholesale loss of Btu's and to encourage the wood's complete combustion. A firebox with a brick or concrete block surround lined with firebrick or a cast able refractory material (a kind of fire-resistant cement) will help to

Considerations for Heat Source Selection						
Heat Source	Heat Value (dry), Btu/lb.	Form	Special Equipment	Boiler Type	Advantages	Disadvantages
Agricultural residues	3,000 - 8,000	Solid	Handling, collection	Batch burner fire tube Fluidized bed	Inexpensive, on-farm	Low bulk density requires large storage capacity
Coal	9,000 - 12,000	Solid	Smokestack scrubber	Crate fire tube Fluidized bed	Available technology	Air quality issues, cost
Wood	5,000 - 12,000	Solid	Chipper or feeder	Fluidized bed	Low cost, clean-burning	Not uniformly available
Municipal solid waste	8,000	Solid	Sorting	Fire tube Fluidized bed	Inexpensive	Limited to urban applications
Geothermal	N/A	Steam, hot water	Heat exchanger	Water tube	No fuel cost	High initial cost
Solar	N/A	Radiation	Thermal collectors	Water tube	No fuel cost	High initial cost
Wind	N/A	Kinetic	Turbine, batteries, heat sink	Electric	No fuel cost	High initial cost

Note: Adapted from Fuel From Farm, US Dept. of Energy

retain a large portion of the heat generated. One of the disadvantages of wood heat is that wood is not a uniform fuel source and temperatures can sometimes be tricky to control with any real accuracy.

Wood's moisture content will vary if it's not dried with

Photo from: R. F. (2009). *Alcohol Fuel A Guide to Making and Using Ethanol as a Renewable Fuel.* Canada, New York: New Society.

consistency, which changes the fuel's Btu output. With a bit of experience, you can learn to control the fire's intensity by adjusting the amount of air admitted through the damper or air inlet.

Aspiration can either be natural, or preferably blower-driven for improved combustion. You can buy universal cast-iron doors with integral inlets, or fabricate your own air controls using sections of black pipe and slotted caps. Be aware that doors made from angle iron and plate steel will eventually warp in the presence of high heat, so you'll either have to protect them with refractory on the inside or use cast iron.

6.6   Co-generation

I want to mention another source of heat energy before moving on to alcohol by-products.

Cogeneration is the application of waste heat from one operation to provide heat for a secondary operation. I already talked about the use of steam above, but there are other energy sources on a far smaller scale that are also fair game. A diesel or gasoline-powered generator or pump creates a lot of heat when running.

Actually, as much as one-third of the energy introduced to an engine as fuel is wasted as heat through the coolant in the radiator. This heat can be directed, very simply, to supply secondary heat to water preheater or to a liquid-to-air heat exchanger to dry residues or calcium salts, or to provide space heat. You may not have a pump or generator on your site, but if an internal-combustion engine is any part of your operation, it shouldn't be overlooked.

6.7   Using By-products

The stillage or leftover mash from an ethanol run can be fed to farm animals as a protein supplement. These distiller's feeds are essentially stripped of alcohol content but retain concentrated nutrients that include proteins, vitamins, minerals, fats and yeast formed during fermentation. The removal of starch and carbohydrates in making alcohol leaves the digestible nutrients at a concentration three times stronger than they were in their original form. So, with corn, the percentage of protein is normally about 9 percent, but increases to 27 percent in the stillage left after ethanol production.

There are four types of distiller's feeds: distiller's dried grains (DDG), distiller's soluble (DS), distiller's dried grains with soluble (DDGS), and condensed distiller's soluble (CDS). The DDG are the solids separated from the mash after distillation. Soluble are the water-soluble nutrients and very fine particles that cannot be easily separated. The condensed soluble and a dried form of these are made from the liquid soluble in the stillage. Feed can also be produced as wet distiller's grains (WDG) to save energy, but usually the solids are separated out using a screen, press or centrifuge of some type and then dried for storage. Moisture content will determine the practical length of storage time, but evaporative drying is costly, so distiller's feeds are often used soon after being processed.

Mature cattle can consume about 7 pounds of dry stillage per day; it should provide no more than 30 percent of the animal's total feed ration in a mix. Other livestock such as poultry, sheep and swine can also benefit from dried distiller's grains on a more limited basis. A local agricultural university or your county or state extension service can provide additional information on feed supplements and particular livestock needs. Carbon dioxide is the noncombustible and colorless gas that we humans and all animals form when we exhale after breathing in oxygen.

During fermentation, yeast manufacture quite a bit of carbon dioxide before they get on to the business of making alcohol. In most cases, on a small scale at least, this CO_2 is simply bubbled off through a fermentation lock and allowed to escape into the atmosphere. But carbon dioxide has value in many types of businesses and industries. It's used in the bottling industry to provide the fizz in soft drinks; it's a benign food preservative and food packaging agent; it's a source of pressure for fire extinguishers, spray paint, and foams or sealants. And, because plants inhale carbon dioxide and exhale oxygen, it's valuable as a crop and greenhouse plant enrichment and growth tool. CO_2, or "greenhouse gas," is actually quite beneficial to the growing industry and is only an environmental problem when it supplants oxygen in the air we breathe.

Analysis of Distiller's Dried Grain Feeds: Corn				
	Distiller's Dried Grains	Distiller's Dried Solubles	Distiller's Dried Grains w/ Solubles	Condensed Distiller's Solubles
Moisture, percent	7.5	4.5	9.0	55
Protein, percent	27.0	28.5	27.0	25.4
Fat, percent	7.6	9.0	8.0	20.0
Fiber, percent	12.8	4.0	8.5	1.4
Ash, percent	2.0	7.0	4.5	7.8
Total Digestible Nutrients, percent	83	80	82	92

Note: Sourced from Distillers Feed Research Council, Cincinnati, Ohio and *Angus Journal*, July 2007.

Photo from: **R. F. (2009).** *Alcohol Fuel A Guide to Making and Using Ethanol as a Renewable Fuel.* **Canada, New York: New Society.**

Realistically, many small alcohol plants cannot generate enough CO_2 to justify the investment in the scrubbers, compressors and storage containers needed to market the gas on a wide scale. But if there were a greenhouse on site or a grower nearby open to an arrangement for putting scrubbed gas to use (CO_2 should be washed of trace elements from distillation), it may be worth going through this simple step to get rid of it in a useful way.

Before leaving the by-products discussion, it's important that you understand that stillage is truly a key co-product to distillation. Depending on how much effort you want to contribute to the operation, it also can be used as a fertilizer, it can be composted, and it can even be used as a feedstock in methane production, which opens up an entire world of possibilities with regard to generating energy for cookers and still re-boilers. Methane constitutes about 94 percent of what we know as natural gas, so with some minimal processing (removal of caustic sulfur dioxide and subsequent drying) it makes an excellent burner fuel.

R. F. (2009). *Alcohol Fuel A Guide to Making and Using Ethanol as a Renewable Fuel*. Canada, New York: New Society. **Retrieve** from: www.newsociety.com

7 Fermentation and distillation

equipment

Economically, putting together the right equipment can mean a lot of difference in the fuel production. Well you can choose to buy all your equipment which might be costly or you choose to build your equipment for the fuel ethanol. In this chapter we will look at the individual component needed for the assembly of the distillery.

7.1 GRINDERS AND SHREDDERS

I've already mentioned grinders, shredders, Hammer mills and presses in the previous chapter, but they're worth returning to because just about any feedstock you're apt to make use of will require some pre-processing. Moreover, some crops (grains) are more suited to milling than fruit is which is more appropriately shredded to pulp. For someone just getting started in a micro sized operation, grains can be milled at a feed mill for a lot less cost than purchasing equipment up front. I wouldn't recommend this as a long-term practice because processing costs can really run up over time, but as a starter it's probably a good choice. If you're ready to purchase equipment, consider scale first. Unless you can get a hammer mill at a very low price (not likely) there's no point in buying one right away if you're not sure how serious you are about producing alcohol.

A **hammer mill** is a bulky, heavy-duty piece of equipment that uses spinning, swinging hammers to crush grain and other materials against a curved steel plate in close tolerances. For grain processing, the plate is perforated as a screen, and different screen sizes can be selected by changing the plate. A hammer mill is a versatile implement, but it does not always deliver a consistent grind. Used construction and agricultural equipment dealers can be a good source of hammer mills.

Shredders are more common and are available even in the economy range, suitable for weekend gardeners and homeowners. Pay particular attention to the design when scouting for shredders, because some use a drum and knife system while others (generally the larger capacity ones) borrow from the hammer mill design. Smaller shredders or choppers used for mulching leaves and branches generally have four-stroke gasoline engines in the 3.5 to 5- horsepower range. They can be built cheaply, with thin-gauge metal hoppers and economy bearings, so overall mass is a good measure of durability. As you move into the 8 to 12-horsepower range, quality improves with the use of large, sturdy hoppers and substantial shafts and bearings. Shredders can handle a wide variety of feedstocks, both dry and liquid, and they can process a good quantity of product if they're large enough. Landscaping supply houses, used equipment vendors, and occasionally equipment rental outfits are all fair game as a source for shredders. It's a good idea to make certain that replacement parts are readily available from the manufacturer for the equipment you end up choosing.

Roller presses and hydraulic or screw presses are more or less specialized for certain types of crops, or they must be used in conjunction with shredding or chopping to extract juices. For grain crops only, there are also gristmills— some with substantial capacity—

that can be purchased from mail-order food supply houses or from agricultural out lets. **Gristmills** are very consistent in the grind they produce and would be a good choice for someone with a fixed source of grain feedstock.

7.2   DEWATERING APPARATUS

Prior to drying on racks or other means, wet distiller's grains or other feedstock substrates may need to be dewatered. There is commercial- scale dewatering equipment available, but small-scale producers would be better off economically by considering some simpler options or modifying other types of equipment for the purpose.

A **hydraulic or simple screw press**, such as the kind used to make apple cider, is a good starting point. By up scaling the design and using sturdier frame and basket components, it's possible to come away with a relatively inexpensive water-removing device. A standing hydraulic arbor press in the 12 to 25-ton range is a good example of the kind of appropriate technology that works well for livestock feed processing. Stainless steel washer tubs from commercial clothes washers or dryers are ready-made baskets for the clever fabricator.

Although it requires more investment, a used auger press makes an excellent dewatered. This design uses a heavy, rotating auger mounted inside a sturdy perforated cylinder. The worm driving force of the auger compresses the mash against a port at the rear of the cylinder, which can be adjusted to release at a predetermined pressure. Until that happens, water is pressed out of the material and is forced through the holes in the cylinder. Auger presses are used in many types of material handling and should be available from a used equipment broker.

7.3   Tanks

Determining the right size for your needs may be a matter of economics. Not all that long ago, when small dairy operations were routinely going out of business all across the country, used stainless steel tanks and vessels were readily available at very reasonable prices. Depending on their function, many, such as milk coolers, were equipped with agitators and cooling jackets. With the passage of time, developing economies in China and India sourced a tremendous amount of scrap steel from salvage yards in the US and Canada, including plumbing, tanks, tractors and agricultural equipment, much of which was questionably "scrap" but sold by weight nonetheless.

The supply of reasonably priced large agricultural and industrial handling equipment has subsequently dried up, but much more recently, global economic slowdowns have pulled the bottom out of the scrap steel market, so salvage tank prices should be shifting to a buyer's market, even at the individual level. Size should actually be determined by how much volume you're prepared to distill at one time. For a 6- or 8-inch column, appropriate for running a batch economically in a time frame between 4½ to 8 hours, a 750 to 1,500-gallon vessel will be sufficient. For the cost-conscious, a 6-inch column mounted on a 500-gallon tank is an ideal starter setup in terms of initial investment and manageability.

The capacity of various sized tanks is illustrated in Table 8.1, which shows ranges from just below 750 gallons to over 5,000.Mild steel tanks are acceptable given reasonably thick wall material. Stainless steel is excellent for its longevity, but stainless is usually costly and requires special skills and welding equipment to fabricate modifications. If you intend to use a single tank for cooking, fermenting and distillation, it's important to plan for some additional capacity — perhaps 20 percent or so—for expansion and foaming in the cooking and fermenting stages.

Some consideration should be given to the vessel's configuration with regard to heating, as well. The more surface area exposed to flame, the more efficiently and quickly its contents will heat. Hence, for a wood-fired setup, a horizontal tank design would be more desirable. An oil- or gas-fired system can concentrate heat more effectively, so a vertical tank may work for fuel oil or gas-heated systems. Larger distilleries often use steam as a heat source, and it is injected directly into the mash or routed through strategically placed tubing, so the tank's configuration isn't as important.

Another practical consideration is access to the inside of the tank for cleanout and maintenance. If possible, choose a vessel with a large enough access hatch to allow human entry, or at least get one that will allow you to fabricate or install a purchased hatch head. A

16 by 16- inch square or 20-inch round head should be adequate, and in any case the lid should be equipped with a positive latch, and it should be well sealed. For a cooker especially, it is very difficult to clean the inside properly without physically scraping any burned residue from the surfaces. If you're cooking in a separate vat and pumping mash to the tank for fermenting and distillation, the human-sized access may not be as critical. Remember, too, that you may have to modify the tank for your own plumbing needs — for example you might want to add internal heat exchangers, enlarge plumbing and drain ports to suit your transfer pumps, or include modular agitators for mixing.

	Vertical Tank Capacities, US Gallons						
Depth, ft.	Diameter, ft.						
	2	3	4	5	6	7	8
1	24	53	94	147	212	288	376
2	47	105	188	293	423	575	752
3	70	158	284	440	634	863	1128
4	94	211	376	587	846	1151	1504
5	117	264	470	734	1058	1439	1880
6	141	317	564	881	1269	1727	2256
7	164	370	658	1028	1481	2015	2632
8	188	423	752	1175	1692	2303	3008
9	211	475	846	1322	1904	2591	3384
10	235	528	940	1469	2115	2879	3760
11	258	581	1034	1616	2327	3167	4136
12	282	634	1128	1763	2538	3455	4512
13	305	687	1222	1909	2750	3742	4888
14	329	740	1316	2056	2961	4030	5264
15	352	793	1410	2203	3173	4318	5640

Photo from:R. F. (2009). *Alcohol Fuel A Guide to Making and Using Ethanol as a Renewable Fuel.* Canada, New York: New Society.

7.5   Pumps and Plumbing

There is little so discouraging as having to handlelarge volumes of warm, fermenting liquidswith 5-gallon pails in an operation that dependson maintaining a schedule. So, pumps and thecorresponding plumbing that go with themare a very important part of your plan. As withmost tools and equipment, you can easily getcarried away with purchases, so unless you'reworking with a liberal budget, you can plan onpumps doing double-duty where possible andusing flexible hoses with quick-disconnect fittingsin addition to hard plumbing. If you plancarefully, you'll probably be able to use gate valves and iron pipe to isolate one delivery orreturn zone and initiate another using thesame pump.

Fittings and valves can be expensivein larger sizes, though, so work out acomparison between the cost of additionalpumps and the cost of hardware before youjump in with both feet.Though positive displacement pumps aregenerally acceptable for distillers, the gear typedoes not work well with mash slurry becausethe particles become bound in the gear teeth,damaging them or the motor. Rotary screw pumps handle slurry far more effectively.

Centrifugal pumps are a good choice, but have the disadvantage of having to be "wet" to start effectively (they're difficult to prime if they're not located at the lowest point in the plumbing) and they may have problems moving thick liquids unless they have a large capacity.

Another type of pump worth considering is the diaphragm style, which uses a tough, flexible skin to expand and contract the pump chamber, thus drawing in, and then pushing out, the material to be transferred. These pumps are sturdy, nearly clog-free, and their design is such that they deliver a consistent volume. With any choice you make, be sure the equipment is able to handle hot liquids and the delivery rate you require. Very small trial plants may get by with modest equipment, but serious operations will need to be moving 30 or more gallons per minute at 40 psi, at heights of 20 feet or so.

When laying out your operation, remember that your pumps will be handling everything from water to liquid slurry to nearly pure alcohol. Potentially, you'll need pumps for feedstock delivery, mash cooking agitation, mash transfer to fermenting and distillation tanks, condenser and heat exchanger water delivery and return, mash feed and stripper column or stillage return (in a continuous design), rectifier condensate return (in a continuous design), reflux delivery (in a direct-reflux design), alcohol storage, and cleaning-disinfecting solution, for which you'll want a separate tank as well. If you use hot oil as a transfer medium, you'll also need a suitable pump for that.

The decision to use rigid pipe or flexible hose may come down to a matter of cost, but keep in mind that it's far easier to remove and clean—or clear out—flexible hose than it is solid threaded pipe. Whatever you choose, it will have to be cleansed of bacteria on a regular basis to prevent infection of the entire system it's connected to. Speaking of connections, threaded fittings may not remain leak-free if conventional pipe sealants are used, especially once your ferment is distilled. The most available and economical choice here is to use Teflon tape and Teflon pipe compound, as it stands up well to the effects of ethanol in varying strengths.

7.5 The Cooker Vat

The mash cooker can be a separate vessel, or, in the interest of economy, it can double as a fermenting vat and even serve as the distillation boiler tank as well. This piece of equipment is as important as the column itself, because thorough cooking is the key to successful starch-to-sugar conversion. If conversion is inadequate, a significant portion of the feedstock is wasted to inefficiency, and the cost of producing alcohol rises considerably.

7.6 Distillation Columns

The distillation column can be made from a variety of materials, as long as it can be effectively sealed and is able to withstand temperatures in the range of boiling water vapors. Commercial columns are commonly made of stainless steel, but mild steel pipe or thin wall tubing is more readily accessible, easy to work with, and available in many standard diameters. Small experimental-type stills can even use 3-inch polyvinyl chloride (PVC) pipe for column material.

Column diameter determines the production capacity of the still, measured on a per-hour basis. Conventional practice is to match the size of the column to the tank capacity, based on how long you want each run to extend, in hours. A comfortable run would be the equivalent of a regular work shift, somewhere between six and ten hours. This isn't necessarily carved in stone, but there's little reason to vary the accepted formula. Too small a column on a large-capacity tank makes for an overly long run with the potential for extended fuel consumption. A large column on a small boiler makes it difficult to achieve equilibrium and deliver the volume of vapor needed to maintain it.

Doubling the size of the column increases its capacity by a factor of four. The table below indicates the contents of various sized pipes and cylinders as well. If you're using salvaged or homemade materials, the cost of a larger diameter column may be worth the investment in terms of output, but if you're purchasing commercial materials especially packing; it may get expensive. The next section will discuss packing specifically, but no salvaged substitute can quite duplicate the performance of well-designed commercial packing.

Volume of Pipes and Cylinders		
Diameter, In.	Volume, per Foot of Length Cubic Feet	US Gallons
3	.0491	.3672
4	.0873	.6528
5	.1364	1.020
6	.1963	1.469
7	.2673	1.999
8	.3491	2.611
9	.4418	3.305
10	.5454	4.080
11	.6600	4.937
12	.7854	5.875
13	.9218	6.895
14	1.069	7.997
15	1.227	9.180
16	1.396	10.44
17	1.576	11.79
18	1.767	13.22
19	1.969	14.73
20	2.182	16.32
21	2.405	17.99
22	2.640	19.75
23	2.885	21.58
24	3.142	23.50

Photo from: R. F. (2009). *Alcohol Fuel A Guide to Making and Using Ethanol as a Renewable Fuel*. Canada, New York: New Society.

While we're on the subject of performance, now is a good time to discuss column height. The height of the column has a direct influence on proof yield of the distillate, because the greater the height, the more water can be stripped from the alcohol or water vapor mix. There is a point, at around 192 proof, where no more water can be removed by distillation no matter how tall the column. But, by means of meticulous heat control and use of the right packing materials, column height can be reduced slightly while still maintaining a high yield. The rule of thumb is a height-to diameter ratio of 24:1 for a 190-proof product. In other words, for every inch of diameter there should be 24 inches of height; a 6-inch column would therefore be 12 feet tall, and a 12-inch column 24 feet in height. Table below shows some design criteria for batch-packed columns between 6 and 15 inches in diameter with bottom vapor entry.

Design Criteria for a Packed Column												
Column Diameter, In.	Height for 180 proof, Ft.	Height for 190 proof, Ft.	Pall Ring Sz. plastic, In.	Rating	Pall Ring Sz. metal, In.	Rating	Saddle ceramic, In.	Rating	Saddle metal, Mm.	Rating	Saddle plastic, In.	Rating
6	10	10-12	½	GD	¼	GD	¾	FR	..	..	..	..
8	10	10-12	¾	GD	½	GD	¼	FR	..	..	1	EX
10	12	12-14	1	GD	1	GD	1	FR	25	EX	1	EX
12	12	12-14	1	GD	1	GD	1	FR	25	EX	1	EX
15	12	12-14	1	GD	1	GD	1	FR	25	EX	1	EX
Rating Key: EX (Excellent) GD (Good) FR (Fair)												

Photo from: R. F. (2009). *Alcohol Fuel A Guide to Making and Using Ethanol as a Renewable Fuel.* Canada, New York: New Society.

Small-scale producers will find that insulating the column (and the introductory conduit, if used) helps to conserve heat control fluctuations in temperature. Even a modest layer of fiberglass or a closed-cell foam jacket provides a welcome degree of consistency. Be sure to choose materials that can withstand the temperatures present at the outer surface of the column. The insulation can be secured with wire or cable ties; nothing more permanent is needed.

7.7   Condensers and Heat Exchangers

Technically, condensers and heat exchangers in the column share similar characteristics in that they both present a cool surface designed to come into contact with a hot medium. In the case of the condenser, its purpose is to reduce ethanol from its vapor phase to a liquid phase at the very top of the column — in other words, to condense the distillate vapors so they can be collected in a jacket surrounding the cool element and piped to a storage tank. Heat exchangers also cool vapors within the column, but they're used as means of heat control to keep temperatures at a specific level in one location. Still other types of heat exchangers or jackets add heat to a boiler or cooking vat to raise temperatures.

The elementary condenser of moonshine lore was just a coil of copper tubing set into a cold water bath. The alcohol distillate condensed within the tubing and was collected from the end. This arrangement was problematic in that mineral and protein deposits would eventually clog the line, causing hazardous backpressure in the column and boiler, often causing an explosion with consequent injuries.

Modern condensers are much safer in that the cold water is run through the tubing and the condensate forms outside it in a larger foul-proof housing. Several feet of ¼-inch soft copper tubing is coiled around a section of pipe, which is used as a form, then the pipe is removed from the coil. The coil is then plumbed through the walls of the housing and sealed at the entrance an d exit points. An even more efficient design is the spiral counter flow type, in which the alcohol vapors are sent through a copper tube with a diameter large enough to discourage clogging. The tube is wrapped with copper wire or strip in a spiral pattern, which promotes a swirling action in the cool water passing through a jacket surrounding the tube. This assures more contact and thus greater heat exchange than a conventional condenser.

A variation on the coil condenser is the Liebig condenser, in which the pipe carrying alcohol vapors is wrapped in a larger pipe jacket that contains the cold water. Though simpler, it is not quite as effective as the coil type because it has less contact surface area and, if made of thin wall tubing or pipe, does not offer the excellent heat-transfer attributes of copper.

Cold-coil heat exchangers within the column are also made of soft copper tubing formed around some type of cylinder. The coils in the column are larger in that more linear footage of tubing is used, along with a larger-diameter 3/8-inch tube. Once tightly coiled, the wraps of tubing should be drawn out and expanded, then re-arranged without kinking to fill the entire cross-section of column and give maximum exposure to the internal vapors. The heat exchangers are anchored to the column through ½-inch couplings welded through the column walls. Brass ½-inch pipe adapters and 3/8-inch compression fittings secure the tubing to the couplings.

7.8　The Re-boiler

A re-boiler is a chamber at the bottom of the stripper column that heats the column but allows boiler vapor to be introduced at the midpoint of the column instead of at the bottom. The advantage to this arrangement is that the entire column gets preheated and stabilized before any alcohol is introduced to it, but more significantly, it allows the initial alcohol-rich mash vapor mixture to ascend the rectifier or upper portion of the column without having to work its way through the lower part of the column. It's also a simple way of avoiding having to separate solids from the liquid mash to prevent clogging the packing in the column.

When mash is fed into the column as a liquid, it usually has to be boosted to a minimum of 150°F before being introduced to the column. The preheated vapors eliminate this step and help the column come into equilibrium much more quickly than it normally would. Re-boilers are normally used for continuous distillation, where the rate of mash feed and the volume of liquid in the re-boiler are coordinated, based on the rate of alcohol production in the column. In stills with separate stripper and rectifier columns, the volume of rectifier condensate is controlled in the system as well.

But some batch stills use re-boilers too, for better heat control. Elementary designs heat the re-boiler by submerging it in the hot mash within the boiler tank (remember that the re-boiler is sealed at the bottom, so no mash can enter unless it comes from the column above).More elaborate designs expose the bottom of the re-boiler to direct heat from the firebox in a wood-fired system, or they use a designated controllable heat source such as a gas burner or an electric element. Using mash itself as a heat source is somewhat problematic in that its alcohol content is constantly changing during distillation, altering in turn the temperature at which the mixture boils due to the differential boiling points of ethanol and water.

7.9 Refluxing

Refluxing is the process of returning a portion of the condensed alcohol to the top part of the rectifying column. It's really a means to control temperature at that point, but it also has the benefit of keeping the upper packing or top plates moist with liquid to encourage fractionation. Technically, refluxing refers to direct refluxing, e.g., physically spraying or pumping liquid ethanol into the top of the column. The cooled alcohol is dispersed through a perforated annular ring, where it mists and dribbles down through the packing. But for our purposes, we'll include the condensation of alcohol vapor by an internal heat exchanger as refluxing, as well. The heat exchanger is located at the top of the column, and though it does consume a significant amount of water (which you should cool and reuse, or recycle hot in some way), it saves having to buy a separate alcohol reflux pump and controls. The vapor temperature at the top of the column will be your indication that things are going well ... or not. If it maintains a steady state and vapor is indeed passing through to the condenser, all is well. Once the temperature starts to rise, you'll have to cool the exchanger coils (or add more alcohol in a direct reflux system) until it stabilizes again. If the temperature drops, you'll need to cut back on the cooling water (or alcohol) accordingly.

7.10 Operating a Batch Still

A batch-run still with a packed column is the easiest type of distillation equipment to operate because there is very little to go wrong, given a healthy supply of mash. If you do miscalculate or make an error in heat control, you'll be able to start again, without losing anything but time. Experience will be the best teacher, of course, but here are some general procedural guidelines to help in getting your operation up and running.

Preparing the Tank

With some stills, depending upon their heating method and column design, the mash can be run complete with solids. However, the simpler packed-column stills do not handle solids well because the particulates either clog the packing, or more likely, get scorched at the bottom of the tank in heating. Unless you have an internal agitator or some other method of keeping the solid particles in suspension while the liquid's being heated, you'll need to separate the solid distiller's grains from the mash to create a liquid beer that's in the range of 10 percent alcohol but free from particulate material. Dewatering devices were addressed earlier in this chapter, but for your early "getting used to it" runs, straining, pressing, or other simple and inexpensive means of separating the liquid from the solids is adequate.

Before filling the tank, all valves and cleanouts should be closed and sealed. If the boiler tank is separate from your cooking vat, be certain that the tank is clean and free of residue.

Initiating and Controlling Heat

Start up your heat source. With a gas or electricburner, you'll have more finite control overtemperature, but even with wood you'll be ableto—and, in fact, will have to—control temperaturesin order to run the still for capacityor higher yield. What's the difference? It is simplythat in the lower heat ranges the liquid beerwill merely simmer, keeping the vapor temperaturelow and allowing vapors to rise slowlyand predictably through the column. Withhigher heat, the beer will boil vigorously andthe vapors will rise into the column accordingly,so you'll have to maintain a close watch on the column's heat exchangers to keep it inequilibrium, or your proof level will suffer.

Not surprisingly, the higher boiling rate requires sending a greater volume of water through the heat exchangers to draw away excess heat, though it does increase the still's capacity. Low heating conserves water and keeps proof levels up, but distillation takes longer. If heat input is *too* low, however, the alcohol will not vaporize in the column and therefore can't be condensed to liquid, so production will halt until temperatures are returned to a working level. At the other end of the scale, from both an economic and environmental standpoint, you should make every attempt to recycle or put to use the warm discharge water from the heat exchangers. It's very suitable for preheating the next batch of mash or a domestic hot water supply, or for watering livestock in an agricultural environment.

Maintaining Column Equilibrium

Temperature at the midpoint of the column should be maintained at approximately 185°F, the boiling point of a 50 percent equal mixture of ethanol and water. Depending on the design of the column and the packing, the exact ratio at midpoint may vary, so the temperature may have to be varied slightly as well. The temperature at the top of the column needs to be maintained at a level just above the vaporization point of nearly pure alcohol, which is 173°F.

With a batch still, however, the concentration of alcohol in the beer decreases as ethanol is drawn off further into the run, so the boiling point of the beer will rise and will continue to do so until the mixture is nearly all water and boils at 212°F. To keep the column in equilibrium, you should find a temperature that works with your design and try to maintain it throughout the run. Do not alter flow through the heat exchangers abruptly, as this will swing temperatures too far beyond your target. It's better to make minute adjustments and allow time for them to take effect before making a second or larger adjustment. It only requires a slight opening of a water valve to increase the flow of cold water, but the water must travel through many feet of tubing, conducting heat from the column the entire way, before it can establish a new equilibrium. The thermometer or temperature gauge at the column will indicate the results of your changes and it should be monitored carefully. Too much heat (or not enough cooling) in the column will yield a low-proof product. Too much cooling in the column lowers temperatures below the vaporization point of the alcohol, and the vapors will condense at the heat exchanger coil and fall back down the stripper section and into the boiler tank (or re-boiler, if equipped), never getting an opportunity to rise past the cool coil. More sophisticated designs use automatic flow valves with thermocouple sensors to control cooling flow, but any temperature changes still have to be made incrementally.

Keeping a watchful eye on the proof of the condensate collected from the condenser will help you to establish where temperatures need to be to make a run consistent and productive, and with each run you'll gain more experience.

Finishing Off and Low Wines

The last quarter of the run will show some marked differences from the first three-quarters. As I mentioned earlier, the boiling point of the mixture in the tank will rise as more alcohol is evaporated, condensed and drawn off in collection. Near the end of the run, temperatures in the boiler will be close to 210°F, and maintaining a temperature of even 185°F at the top of the column becomes more difficult. A lot of cooling water will be consumed to remove water from the mixture, and it will take as long to draw off a high-proof product from the final quarter of the run as it did to get the same or better product from the first three-quarters.

From an energy and time perspective, the best course of action at this point is to allow the column temperature to rise and do your best to maintain the midpoint column temperature several degrees below that of the boiler temperature, and the top column temperature several degrees below what the midpoint is. This will allow a portion of the mixture to reflux, or separate into water and ethanol, and will expedite the process significantly, saving valuable heat energy. The proof levels in this final stage will be relatively low, but there is enough alcohol in the product to make it worth saving. These so-called "low wines" can be stored in a separate vessel and combined with the low wines from successive runs to make a whole new distillation run in the future. As with the main distillation run, you will also create a substantial amount of hot water, which should be reused in a productive manner.

7.11   Vacuum Distillation

Under normal conditions, at sea level, there are about 14.7 pounds per square inch of air pressure bearing down on the earth's surface. At higher elevations, there is less air available to press down, so the pressure isn't as great. It makes a difference: As anyone who's lived at mile-high elevations can attest to, liquid boils at a lower temperature than it does in the flatlands, and that can create problems in cooking certain foods.

Distillation in a vacuum, in which air pressure is ma atmospheric pressure, takes advantage of this fact. When distilling alcohol, we're not necessarily seeking a particular temperature. Our goal is to get the liquid boiling with as little energy input as possible (in heat or Btu's). So, vacuum distillation offers the advantage of energy conservation, plus — once the system is brought into equilibrium — it's easier to maintain in a steady state.

Let's look at this in detail, because distillation in a vacuum isn't as simple as it might appear, nor is it the solution to all ethanol production problems. First, it requires a vacuum pump of some sort. Commercial models used by refrigeration and cooling professionals cost hundreds of dollars—and even thousands for stationary units; compressors salvaged from box or chest freezers cost far less, but their oil seals can be damaged by alcohol vapors in the air. The best choice is oil-less pump capable of drawing at least 0.5 cubic feet per minute and 25 inches of mercury (In. Hg), a measure based on its ability to pull the heavy liquid element up a tube (a perfect vacuum is just over 29.9 In. Hg by gauge measurement). That kind of vacuum pressure can be fatal to thin wall steel, so it's important that the column, tank and plumbing be substantial enough to withstand the stress. At 25 In. Hg, water boils at around 130°F and alcohol at about 106°F.

Water flow to the cooling chamber and condenser at the top of the column must be meticulously controlled to regulate temperature at that point, so a temperature-actuated modulating valve, rather than a manually controlled valve, is recommended. So is an accurate means of measuring temperatures at both the mash boiler and the top of the column. A reliable vacuum gauge for the tank is necessary, as well as a fine-control needle valve to introduce atmospheric air into the system for initial heat-up and to control vacuum if needed.

Finally, from the standpoint of safety, a pressure safety release valve should be installed on the boiler tank of this (and any) distillation vessel to relieve pressure buildup if the vacuum pump should fail. A 15-psi limit is safe; standard water-heater temperature and pressure valves may have too high a pressure threshold (150 psi) and a marginal temperature limit (210°F), especially for atmospheric distillation.

Low-temperature distillation opens the door to solar thermal preheated water and even direct solar thermal water if concentrating collectors are used. Other forms of renewable or biomass-based energy might be practical here, where they would not be otherwise—for example, an electric-element heat dump from a wind or micro hydro turbine, or co-generated heat. Lower temperatures also pose less operating risk from accidental burns, which can be serious at normal boiling levels. Vacuum distillation, on the other hand, comes with reduced capacity unless you opt for a continuous distillation process.

You can expect up to 40 percent less capacity per hour for a given size batch column under vacuum, meaning that initial material costs will be greater from the outset to achieve equivalent yields. Continuous distillation means that you'll have to filter your mash before introducing it to the column, because the packing easily becomes clogged with solids. Controlled cooling under vacuum also requires a substantial amount of water because the temperature of the vapors at the top of the column isn't that high to begin with, so the temperature differential is narrow. Either a lot of water flow or very cool water input is needed to draw down the temperature to working levels.

Operating a vacuum still involves, by and large, getting the column into equilibrium. Firing the re-boiler with water in a normal atmosphere allows it to build up heat to the point where it will boil at reduced atmospheric pressure. When that happens (at a temperature around 130°F), the vacuum pump can be activated and the atmosphere drawn down to or near the desired level of 25 In. Hg, at which point mash can be introduced to the column at midpoint. From here, heat control through the re-boiler and via the top coil is critical. Too much heat in the re-boiler overheats the column and delivers a low-proof distillate. Too much cooling at the top causes premature condensation and prevents alcohol vapors from passing into the condenser; they simply reflux and fall back down the column. Conversely, not enough cooling at the top creates the equivalent of an overheated column. The vapor mixture will not fractionate and low proof will result. The mash feed rate must also be coordinated to optimize with the rest of the system or the column will flood and fractionation will not occur. The question of whether vacuum distillation is worth the investment expense goes back to how "appropriate" you want your distillery to be. If your fuel costs are high and you're running a continuous process on a regular or frequent schedule, then an artificial atmosphere still may have some cost-saving value to you, even given the additional initial expense. If you're a casual fuel producer operating a batch still on the scheduled occasions when you need fuel, vacuum technology makes considerably less sense.

7.12 Alcohol Storage

Bulk storage after the fuel has been denatured can be above or below ground. If you feel that a standing tank might be a target for theft, it can be protected with a simple enclosure such as a panel or chain-link fence. Underground storage has the benefit of being out of sight and out of the way. Unfortunately, the incidence of underground leakage, especially from service station tanks containing gasoline with the carcinogenic MTBE additive, has prompted state and federal environmental agencies to regulate buried fuel storage well beyond any practical return, albeit for the protection of our groundwater. Single-wall steel tanks are particularly prone to the corrosive effects of wet soil, and leaks are very difficult to detect deep in the soil.

That leaves above-ground storage, which is actually less of a risk to the environment because any leak or damage is obvious and can be repaired promptly. The fact that 190-proof alcohol has a flashpoint—the lowest temperature at which vapors can ignite in air with a source of ignition — over 100 degrees higher than that of gasoline, (which is still dangerous at minus 45°F) indicates that it is safely stored above ground.

The simplest type of tank is the horizontal cylinder mounted on a steel ladder frame several feet from the ground. The fuel from these tanks is gravity fed through an automatic shutoff nozzle, usually secured through a ball valve on the tank outlet. The frame's four feet should be bolted to substantial concrete pads set in the ground, preferably 16 inches square and excavated to the frost line for your region. Five-hundred gallons of fuel ethanol weighs over 3,300 pounds, enough to sink an unpadded leg deep enough into wet ground to cause the frame to buckle. It would be a good idea to check with your local code jurisdiction to see if any other requirements have to be met, such as earthquake protection or spill containment.

A full-size slab with a short perimeter knee wall may be required. Other than a fuel delivery fitting, the tank should have a 2-inch fitting for the fill cap and drop tube, and another one for a combination pressure-vacuum vent, sometimes called a conservation vent. This is a dual-action device that safely releases excess pressure in the tank in hot weather, and prevents outside air from being sucked into the tank when it cools until a significant vacuum is formed. A mechanical fill-level gauge is also a convenient item. If possible, avoid setting your tank in direct sunlight, or at least provide some type of shade screen or lean-to that is non-combustible. If a frame-mounted tank isn't feasible for aesthetic or safety reasons, you'll have to use a pad-mounted surface tank. The pad should be reinforced concrete and the tank supported off the slab on the short metal feet attached to it. Because you won't have the benefit of gravity here, one of the 2-inch fittings will have to accommodate either a crank-operated or electric suction pump with a tube and pickup strainer.

Besides any grounding associated with the breaker box for an electric pump, the tank itself must be grounded, as well as the filling nozzle, which should be grounded to the tank, unless it's non-metallic. Static electricity can build up in both steel and reinforced plastic tanks and must be sent to earth to prevent any risk of sparks during handling.

Fuel alcohol, even manufactured at a small plant, is not supposed to be stored in containers of less than 5 gallons in size except for labeled testing samples, according to federal TTB regulations. Also, any alcohol fuel container of less than 55 gallons should be designated with a label that states "Warning— Fuel Alcohol — May Be Harmful or Fatal if Swallowed." For safe drum storage, do not store lower-proof alcohol fuel in steel drums, as they can corrode at the seams from the inside. Higher-proof ethanol can be stored in steel drums, but only if they're positioned vertically, with the fittings at the top, as the threads tend to leak.

It's always a good idea to run the fuel through a filter before pumping it into the vehicle tank. Three-micron and even one micron filters are the best choice, as they'll remove even the finest contaminate particles, but be sure you choose an alcohol-tolerant filter; most conventional fuel filters are designed to separate and absorb water and will eventually become plugged, especially with alcohol below 190 proof.

R. F. (2009). *Alcohol Fuel A Guide to Making and Using Ethanol as a Renewable Fuel.* Canada, New York: New Society. **Retrieve** from: www.newsociety.com

8 ALCOHOL as Fuel Engine

For decades, the cheap and easy availability of gasoline has led us to believe that the petroleum-derived fuel is the only one suitable for internal combustion engines. The petroleum industry has dedicated itself to promoting its products, and, using a combination of skillful marketing and successful government lobbying, it has built a massive transportation and industrial infrastructure on derivatives of oil.

Yet, gasoline is only one of several fuels that have proven themselves to work in a gasoline engine, and in fact kerosene, propane, natural gas and ethanol have all at one time or another served as dependable fuel sources. Aside from the limitations of infrastructure (which is a significant reason why alternative fuels can't seem to get traction in the marketplace), a large part of the reluctance to consider fuel alternatives is simply that auto engines have been optimized for gasoline. Changing that is not impossible, and not even difficult, but clearly would require some investment and incentive, which doesn't always come easily.

8.1 Combustion Properties

Gasoline is a complex mixture of hydrocarbons — molecules made up of hydrogen and carbon — and other elements such as sulfur, nitrogen, boron and phosphorus. It is a combination of dozens of different compounds, which vary according to the source of the crude oil from which it's made. When the crude oil is heated and distilled at the refinery, it is separated into different products according to its intended use: petroleum gases and solvents, aviation gasoline, automotive gasoline, kerosene, heating oil, diesel fuel, lubricants and waxes, and furnace oil and asphalt. The compounds that boil at lower temperatures, such as gasoline, are lighter, and those with the higher boiling points—lubricants and furnace oil— are heavier.

There is no "typical" gasoline formula because refineries further vary the blend to suit seasonal changes, altitude, emissions requirements, and so forth. But, for a representative sample, we can say that octane is as good an illustration as any, with its 8 carbon atoms and 18 hydrogen atoms (C8H18). Ethanol, as you've learned, is also distilled from carbohydrates, which contain carbon, hydrogen and oxygen atoms. With ethanol, one of the hydrogen atoms has been replaced with a hydroxyl radical, which is an oxygen atom bonded to a hydrogen atom. The molecular formula for ethanol — C2H5OH — reflects this. Ethanol is only one of many different types of alcohol, but only ethanol and methanol (a toxic alcohol derived from wood distillates or synthesized from natural gas) are normally used as fuel. Methanol is popular in sprint and drag racing, and it was the fuel of choice of the Indy Racing League fromthemid-1960s until 2007, when it was replaced with 100 percent ethanol. But methanol is only mentioned here as a point of discussion—it is not economically feasible to make on a small scale and it is toxic even in small quantities or when absorbed through the skin—hardly the ideal home-shop project.

The differences between gasoline and ethanol are abundant enough to comment on, and not simply because of the oxygen present in ethanol. Table below compares the properties of the two, and in the following few sections I'll point out a few points of particular interest.

Liquid Fuel Characteristics of Gasoline and Ethanol		
Chemical Properties Formula	Gasoline complex mixture	Ethanol
Molecular Weight	variable	46.07
Percent Carbon (by weight)	85-88	52.14
Percent Hydrogen (by weight)	12-15	13.12
Percent Oxygen (by weight)	variable	34.74
Carbon/Hydrogen Ratio	5.6 - 7.4 : 1	4.0 : 1
Stochiometric Ratio (air-fuel)	14.2 - 15.1 : 1	9.0 : 1
Physical Properties	Gasoline	Ethanol
Specific Gravity	.70 - .78	.7936
Liquid Density (Lb./Cu. ft.)	43.6	49.3
Liquid Density (Lb./Gallon)	5.8 - 6.5	6.59
Boiling Point (Degrees F.)	80 - 440	173.3
Freezing Point (Degrees F.)	minus 70	minus 174.6
Solubility (in water)	240 ppm	miscible
Solubility (water in)	88 ppm	miscible
Vapor Pressure at 1 Bar (100°F.)	7 - 15 In. Hg.	2.5 In. Hg.
Vapor Pressure at 1 Bar (77°F.)	.3 In. Hg.	.85 In. Hg.

Liquid Fuel Characteristics of Gasoline and Ethanol		
Thermal Properties Formula	Gasoline complex mixture	Ethanol
Heat of Combustion (77°F.)		
Lower Heating Value (Btu/lb.)	18,900	11,550
Lower Heating Value (Btu/gal.)	115,400	76,114
Higher Heating Value (Btu/lb.)	20,250	12,780
Higher Heating Value (Btu/gal.)	124,800	84,220
Latent Heat of Vaporization (77°F. at 1 Bar) (Btu/lb.) (Btu/gal.)	150 900	395 2,603
Flash Point (Degrees F.)	minus 50	55
Autoignition Temperature (Degrees F.)	430 - 500	793
Octane Rating (Research)	90 - 101	106
Optimum Air-Fuel ratio	15 : 1	9 : 1
Explosive Limits Air-Fuel Ratio	13.2 : 1 - 71.4 : 1	5.3 : 1 - 23.3 : 1
Explosive Limits in Air (by percentage)	1.4 - 7.6	3.3 - 19
Maximum Practical Compression Ratio (spark ignition)	9.2 : 1	15 : 1

Photo from: Photo from: R. F. (2009). *Alcohol Fuel A Guide to Making and Using Ethanol as a Renewable Fuel.* Canada, New York: New Society.

8.2   Emissions

The combustion of fuel in an engine generates by-products that we all know as emissions. The four main automobile emissions are hydrocarbons, carbon monoxide, oxides of nitrogen, and carbon dioxide (though others, such as particulates and formaldehyde, are also produced). It is clear that anthropogenic, or human induced, pollutants are spiraling out of control and must be managed for the sake of our environment and our future well-being. But even from a basic economic viewpoint, a clean burning fuel is a more efficient fuel. It reduces energy waste and extends the life of internal components by decreasing the incidence of damaging carbon residue within the engine.

Gasoline, as a compound hydrocarbon, is not a particularly clean-burning fuel. Every motorist knows the danger of allowing a car to idle in an enclosed space; what may be lesser known is that auto emissions are responsible for 53 percent of the typical US family's annual contribution of CO_2—nearly 27,000 pounds per year.3

Ethanol, in comparison, burns nearly pollution- free. It already contains oxygen integral with the fuel, which can lead to a more homogenous combustion. It burns with a faster flame speed than gasoline, and it does not contain additional elements such as sulfur and phosphorus. All these factors work in ethanol's favor with regard to emissions—as well as the fact that its lower exhaust gas temperatures tend to reduce NOX (nitrogen oxide) emissions specifically. At low load, NOX emissions with ethanol are lower than with gasoline, due to differences in the latent heat of vaporization and in the combustion speed.

The low exhaust temperatures produced when using ethanol enable a balanced combustion mixture along the whole operational range, and make it possible to use a 3-way catalyst to reduce NOX emissions, even at full load. Coming up with accurate emissions data for a "typical" engine is difficult, however, for a number of reasons. Tests have been conducted by a number of universities, the US Department of Energy's Solar Energy Research Institute (SERI) and its Office of Alcohol Fuels (OAF), the Environmental Protection Agency, various automobile manufacturers, and by the petroleum industry itself. The research spans many years, but the tests were conducted with diverse goals. Some concentrated on alcohol gasoline blends; others analyzed the use of methanol; many made modifications to the engines that altered results; some concentrated on engine wear patterns and longevity.

8.3 Performance and Fuel Economy

Fuel economy, or gas mileage, is relative to engine load and the mixture of air and fuel that's burned in the cylinders. Fuel, no matter what kind, must be mixed with air before it will combust completely inside an engine. Whether the fuel is blended with air before it enters the intake manifold (by means of a carburetor) or whether it is mixed in the manifold or directly in the combustion chamber (by means of fuel injection), a certain ratio of air to fuel must be attained in order for the mix to burn efficiently.

Each fuel has its own proportion, known as the stoichiometric ratio. Gasoline needs 14.7 parts of air to each part fuel. Ethanol, because it contains about 30 percent oxygen by weight, only requires 9 parts air to each part alcohol. These figures are theoretically precise, but in the real world, they will vary somewhat depending upon load and speed demands placed upon the engine.

A lean condition in which the balance of air to fuel is greater than normal results in better fuel economy but higher exhaust gas temperatures. As more air (or less fuel) is introduced to the mixture, the engine may temporarily increase rpm, but performance will deteriorate, and if the leaning trend continues, damage to the valve facings will occur as internal temperatures reach a critical point.

Conversely, a rich condition is one in which the balance of fuel to air is greater than normal. Adding more fuel (or less air) to the mix lowers exhaust gas temperatures and increases power output, but also lowers fuel economy and contributes to hydrocarbon emissions. Most every contemporary automobile is equipped with fuel injection, which automatically compensates for load and environmental variables through a system of sensors and computerized controls. Cars, of course, are optimized for gasoline. But piston-driven aircraft can provide some insight into the performance characteristics of ethanol fuel because they operate in varying altitudes and weather conditions and are regularly exposed to extreme changes in air pressure and temperature. In aircraft, the pilot manually adjusts the air-fuel mixture to maintain an ideal ratio while atmospheric pressure and temperatures vary.

In certification testing done on a Lycoming O-235 aircraft engine in 1992, the maximum horsepower achieved by the engine when fueled by 100-octane Low Lead avgas (aviation gas) was approximately 125 horsepower. A 20 percent increase (25 hp) was gained when the fuel was changed to ethanol.

Vehicles with fixed-jet carburetors or simple air-cooled engines don't have the flexibility to optimize ethanol fuel without modifications. But even these fundamental engines can be made to perform well on ethanol with a few basic changes, especially when ethanol is the dedicated fuel. What follows points up a few areas for consideration when contemplating a conversion to ethanol fuel.

8.4   Engine Modifications

In this section, I'll get into the details of converting a gasoline engine to alcohol fuel, and specifically the modifications and adaptations necessary to get the job done. I'm including engine principles in the segment following, particularly to support those new to mechanical applications. Thereafter I'll be addressing key components—carburetion, fuel injection, ignition timing, and cold-starting systems, followed by some specific modifications and adjustments that will make things easier for engines to accommodate the new fuel.

8.5   Principles of Engine Operation

By and large, the most common type of engine used in everything from lawn mowers to passenger cars is the spark-ignition internal combustion engine (ICE). *Spark ignition* because the fuel is ignited by an electrical spark plug (as opposed to the compression ignition used in diesel engines) and *internal combustion* because the fuel burns inside the engine rather than external to it, as in a steam engine.

These kinds of engines are known as reciprocating or piston engines, referring to the up-and-down motion of a piston in a cylinder. The most common combustion sequence used is the four-stroke cycle, which has four stages:

(1) The intake stroke, in which a mixture of fuel and air is drawn into the cylinder through a valve as the piston moves downward;

(2) The compression stroke, in which the piston moves upward, compressing the fuel air charge; (3) The power stroke, which occurs when the spark plug ignites the charge and drives the piston downward;

(4) The exhaust stroke, which forces combusted gases from the cylinder through another valve when the piston moves upward to complete the cycle.

The piston is connected to a crankshaft, which converts this up-and-down motion to the rotary motion of a shaft. Since the piston makes one up-and-down trip for each full rotation of the crank, one complete four stroke cycle yields two crankshaft revolutions. So, an engine turning at 2,000 revolutions per minute completes 1,000 such cycles each minute.

Another, less familiar type of combustion cycle is the two-stroke cycle, commonly used in chainsaws, weed cutters, small motorcycles, and other applications where maximum power with minimum weight is essential. Two stroke engines use ports instead of valves, which require a timed mechanical connection to operate them. These ports are controlled directly by the engine piston, and the engine crankcase (the chamber beneath the piston that houses the crank mechanism) serves as a compression pump and fuel-mixing chamber.

A transfer port allows fuel to move from the crankcase to the combustion chamber, where the spark plug is located. The design is simpler and uses fewer components, making the engine lighter with no loss of power.

In the two-stroke cycle, when the piston is moving downward the exhaust port is opened first, allowing combusted gases from the previous cycle to escape. Shortly thereafter, the piston reaches the bottom of its travel (a position referred to as bottom dead center, or BDC), and the pressurized fuel mixture is forced through the transfer port into the combustion chamber, forcing out, or scavenging, any remaining exhaust fumes.

The second part of the cycle occurs when the piston moves upward, first closing the transfer port, then the exhaust port. The fuel mixture is compressed while the rising piston creates a vacuum that draws a fresh fuel charge into the crankcase through the intake port. When the piston reaches the top of its travel (top dead center, or TDC) the spark plug ignites the fuel mixture, driving the piston downward and initiating the cycle all over again.

In four- and two-stroke engines, fuel and air is mixed precisely in a carburetor before being sent to the combustion chambers; engines with multiple cylinders use an intake manifold to deliver the mixture in equal proportion to each cylinder. (In most contemporary engines, the carburetor has been replaced with a more efficient fuel injection system)

The rate of spark plug firing is precisely timed to the speed of the engine in relation to the position of the piston. This timing can be adjusted within a moderate range to suit fuel and varying driving conditions such as increased load or altitude changes. The term "ignition timing" refers to the point of spark ignition on a 360-degree scale corresponding to the rotation of the crankshaft. This point can typically vary between 5° and 45° before top dead center (BTDC).

8.6 Carburetion

Pop the hood of any modern car, and you're not likely to find a carburetor. But don't be misled—millions of cars, trucks, motorcycles, boats, aircraft, mowers, pumps, generators and tractors that were manufactured with carburetors are still in use and will be for many years to come. Since the earliest days of the internal combustion engine, carburetors have been the mechanism of choice for mixing fuel and air in the appropriate ratio for efficient combustion. As mentioned above, for gasoline, that ratio is about 14.7 parts air to 1 part fuel; for ethanol, the ratio is closer to 9 to 1.

A carburetor includes a small reservoir, or bowl, where liquid fuel pumped from the tank is maintained at a more or less consistent level by an internal float connected to a needle valve. This micro fuel supply feeds several circuits in the carburetor: an idle circuit that allows the engine to run smoothly at minimum rpm, a main circuit that allows cruising and high-speed operation, and, in more sophisticated carburetors, an accelerator pump and a power valve circuit that help in the transition from low to full power. Let's look at each individually.

8.7   Fuel Injection

1990 was a watershed year of sorts in automotive history, because in that year the last carbureted American passenger car, a 5.0 Liter OldsmobileV-8, left the Detroit assembly line. After years of struggling to meet federal emissions standards and fuel economy goals using cobbled-up plumbing and increasingly complex carburetors; Detroit finally acquiesced to the reality that fuel injection's time had come.

Mechanical pressure injectors actually appeared on diesel engines around 1930, and, during the Second World War, they were refined for use in high-performance gasoline powered fighter aircraft to gain horsepower and shed the restraining effects of gravity on liquid fuel. In the 1950s, a few exotic sports cars and competition racers used injectors, but it wasn't until the late 1960s that they became part of regular production, most notably on certain Volkswagen models. The earliest continuous injectors provided additional fuel on demand, either by increasing the fuel flow rate to each injector through a distributor, or increasing the pressure within the fuel line. Each cylinder had its own injector mounted in the vicinity of the intake valve. Later, pulsed injectors were developed that used electric solenoids to open and close a valve on each injector.

With these, fuel delivery was controlled by how long the valve was held open during each pulse cycle. Pulsed fuel delivery has been simplified by the development of common-rail fuel lines— almost like a liquid manifold for fuel to replace the separate lines to each injector. The pulse timing, and duration, or pulse width, is controlled by the same type of microprocessor control unit used to govern the feedback carburetors.

The data collected by the MCU to determine the most appropriate pulse width for conditions includes the amount of oxygen in the exhaust stream, airflow, ambient temperature of the air, temperature of the coolant, and engine rpm, among other things. As engines and computer controls get more sophisticated, even more information is collected and processed to include altitude, pre ignition, engine timing, and coolant, air and exhaust gas temperatures, which are all used to manage various systems within the engine far more efficiently than they were even a few short years ago. Now that we have the basics down, let's look at the two common electronic fuel injection (EFI) configurations.

Establishing the precise point at which the spark plug fires in relationship to the position of the piston is called ignition timing. Many factors, including the shape of the combustion chamber, the heat within the chamber, the speed of the engine, the compression ratio, and the octane rating of the fuel determine what timing setting is best. Timing can be set so the spark fires anywhere from0° to 45° before piston top dead center (BTDC).The more degrees along the rotation of the crankshaft involved, the greater the timing advance is. As the degree setting moves back toward the zero or top dead center (TDC) point, ignition timing is said to be retarded.

There was a time when timing was established mechanically, directly from a gear on the engine crankshaft or camshaft. Engines with multiple cylinders use a distributor connected to this gear to direct high-voltage energy from the ignition coil to the individual spark plugs. At first, a driver-operated lever that was linked to the distributor controlled the timing changes, within the modest range necessary for engines of the day. Later, engine vacuums and centrifugal weights in the distributor more or less automated the process so no human input was needed. Much later, in the era when environmental controls became an important component of engine design, electronically controlled distributors were introduced that took input from engine-mounted sensors and made changes automatically and constantly as needed.

Today's ignition systems have done away with mechanical distributor altogether and rely on the vehicle's microprocessor control unit to read, analyze and act upon information in a highly efficient real-time process. But, back to the conventional distributor a vehicle or engine-driven piece of equipment has specifications that let you know where to set static timing for normal gasoline operation. Static timing refers to the initial setting, at idle speed, before any dynamic timing advance from vacuum or weights comes into play. The distributor housing can be rotated by loosening a clamp at its base in order to advance or retard the timing of the spark in relation to the position of the crankshaft, and by default, the pistons within the cylinders. You'll need timing light to read the degree marks at the crankshaft's harmonic balancer; the light's strobe flash is synchronized with the firing of spark plug No.

Beginning at the factory setting, first remove the hose to the vacuum advance unit and plug the hose with a plug or small dowel section. Turn the distributor to advance the reading at engine idle to about 12° or 15°. Lock down the distributor at this point and reattach the vacuum hose, then drive or operate the engine under load, and at high speeds. If the distributor advance is too great, the engine will ping loudly on acceleration and misfire at higher rpms. Remember, the static setting is, say, 12°, but the dynamic advance built into the distributor can add an additional 15 to 25 degrees. The point is to run the engine with the highest degree of advance allowable without deteriorating performance or damaging the engine with pre-detonation. If audible sounds of pinging or knocking occur, you'll have to retard the timing by a degree or two and re-test. Continue this procedure until the pinging ceases. With some distributors, there isn't enough range built into the housing to allow maximum advance. In such cases, you can lift the distributor housing from the engine block after removing the hold-down clamp and bolt, and rotate the distributor shaft one tooth forward (in the direction of rotation) to gain the advance needed. Then go through the testing procedures and lock down the distributor when finished.

This is a decidedly unscientific method of setting timing, but it functions well in lieu of a knock sensor, an auditory device that is built into modern electronic systems and is available as an aftermarket purchase. Alcohol can burn in a wider range of air-fuel mixtures than\gasoline can, and has a higher octane rating, which allows it to operate safely with much higher compression ratios. Running at the highest allowable advance lets the engine run cooler, attain better fuel mileage, and develop more power under load. High compression, proof strength of the alcohol fuel, and high energy ignition (HEI) systems all affect the degree to which advance can be dialed in.

Electronic ignition systems have the disadvantage of not being as accessible or modifiable from a backyard mechanic's point of view, as the old mechanical distributors. However, with a bit of investment, there are several aftermarket options open to the tinkerer that will co-exist splendidly with alcohol fuel setups. Some electronic ignition kits made for retrofit onto the earliest engines have a wide range of advance, enough to accommodate gasoline and alcohol fuels. There are also electronic devices made for the performance market that allow either manual control of spark timing or use of a programmable path that lets you set the system up for whatever your situation might be with regard to proof strength, engine load, compression ratio, or fuel mixture.

8.8   Cold-Starting Systems

One point consistently raised by alcohol-fuel opponents is the fact that ethanol does not vaporize as easily as gasoline, and needs some assistance in cold temperature start-ups. E-85blends have solved the problem by mixing the ethanol with 15 percent gasoline, which is enough to set off the fuel charge in most winter climates. Those in particularly cold regions get a special winter blend that contains up to15 percent more gasoline to lower the flashpoint even further.

Still, the issue remains that for people using straight alcohol fuel, temperatures below 40°F or so can be a problem. Carbureted engines have the most difficulty because the fuel delivery is not pressurized. Fuel-injected engines, because of their fuel spray, will start at lower temperatures.

The simplest and least expensive way to provide the flash needed for cold starting is to inject a small amount of gasoline into the manifold at startup from a remote reservoir. Some of our earliest conversion projects used a 5-gallon tank and an electric fuel pump to send a spurt of gasoline through a fuel line and metering jet fastened to the air cleaner housing. Cautionary letters and howls of protest from safety advocates pointed out that this was too risky in the event of a backfire through the carburetor (which isn't all that uncommon), so the newer modifications inject the fuel further downstream, into the manifold or even more conveniently, into the positive crankcase ventilation (PCV) hose which leads directly to the manifold and draws a vacuum while the engine is running.

Plumbing this kind of setup is relatively easy. To start you really only need a small canister, about the size of a windshield washer reservoir or lawnmower tank. Mount the tank under the hood, away from the exhaust manifold or other source of direct heat, and route a ¼-inch fuel line to an electric fuel pump bolted to the firewall or other convenient location. Connect another section of fuel line from the pump outlet to a tee fitting in the PCV hose. The tee can simply be 3/8-inch brass with two 3/8-inch hose barb fittings threaded into the holes opposite each other. The branch can be fitted with a third hose barb (or a fuel line fitting if you're using steel line) that's been modified by soldering a small-orifice metering jet onto its end. Make sure the jet's head is smaller than the diameter of the threads and is bonded to the fitting securely, for you do not want it sucked into the manifold if it comes loose. Don't be afraid to drop down to a ¼- inch thread for the branch if needed to utilize a smaller jet. Tee fittings are made with many types of thread combinations.

You don't even need to use an electric pump for the small amount of fuel we're talking about. I have seen foot-operated windshield washer systems converted to gasoline service (make sure the plastic components are compatible with gasoline; many are), and pumps adapted from camp stoves that use white gas or lantern fuel are ready-made for this purpose. Make certain that whatever type of pump you use is self-closing so that fuel is not sucked into the manifold while the engine is running. If you're concerned about safety, tractor and auto supply houses sell commercial diethyl ether injection kits for installation on diesel tractors and trucks. These are very common in cold climates, and the kits include the ether canister, an electronic solenoid valve, an atomizer, a momentary switch, and fuel delivery line. Some even use a thermostatic sensor to open the circuit if the engine block is warm, shutting off current to the valve when ether is not needed. These types are usually wired into the starting circuit of the vehicle's ignition switch. The ether line is plumbed into a hole tapped into the intake manifold, and a measured amount of starting fuel is injected at startup.

Other creative cold-starting methods have used propane torch canisters and solenoid shutoff valves from LP-powered equipment such as generators and forklifts, and glow plugs from diesel tractors to heat alcohol fuel injected at the manifold. Preheating the alcohol in cold weather can be effective, as well.

8.9   Diesel Engine Block Conversions

Before moving on to turbocharging as a means of boosting compression, I want to address the use of converted diesel engines for alcohol fuel use. Rather than trying to build a high-compression block from a stock gasoline engine, this approach takes an already substantial factory built diesel engine and converts it for spark-ignition use. It's not a low-cost conversion by any means, but it immediately resolves the litany of issues to be faced in modifying a gasoline engine for super-high compression.

Essentially, the diesel block is fitted with gasoline-service components. Some engine blocks are cast with an eye toward sharing as many parts between gasoline and diesel as possible, mainly with regard to casting bosses and bolt patterns. If this is the case, the task is easier, but beware of the GM automotive diesels of the late 1970s, which were modified from gasoline blocks and are somewhat problematic in their durability.

In the US and elsewhere, research has been done in this area. Typically, the engine is Alcohol as an Engine Fuel 195de-compressed from a ratio of 21:1 down to a level more compatible with alcohol; spark plugs are fitted into the glow plug ports; and an electronic ignition system is installed. Stock diesel injectors cannot tolerate ethanol without lubricant, so multi-port injection at the intake manifold provides the fuel. Other modifications, mainly to make the sensor readings compatible with the electronic control module, are fairly straightforward. Realistically, this is beyond the ken and probably the budget of the average shade-tree mechanic, but it's encouraging to know that the possibility exists.

8.10   Using Synthetic Oils

As mentioned above, synthetic lubricants are a better choice to use with alcohol fuels than standard petroleum-based oils. While it's true that synthetic oils can cost two or three times more than conventional oil, there are a few good reasons why synthetics are preferred in your crankcase.

For one, synthetic lubricants have unusually high film strength, due to their long-chain chemical structure. In practice, this means that the lubricant film between metal surfaces such as bearings and crankshaft journals is particularly tenacious. Also, the viscosity range in synthetic oils is fairly broad, meaning that the oil remains thin at cold startup, the more easily circulated through galleries and bearing parts, and thickens as the temperature rises, providing protection at operating speeds.

Furthermore, synthetic lubricants are formulated to not degrade easily at high temperatures, extending their lifespan considerably. It's routine for synthetic oil to be left in the crankcase and only the filters changed for periodic maintenance until the oil has run through a half-dozen or more maintenance cycles.

8.11   Alcohol and Diesel Engines

Though spark-ignition engines have been the crux of alcohol-fuel research, compression-ignition diesel engines have not escaped scrutiny, primarily because they are highly efficient and very durable. I talked about diesels briefly, a few sections earlier, in pointing out that they have extremely high compression ratios, which are essentially what heats, and ignites, the fuel. There are several immediate problems with using ethanol as a fuel in diesel engines. First, the anti-knock qualities that are so attractive in alcohol are actually a drawback when compression is igniting the fuel charge. Diesel fuel has a cetane rating, which is used to gauge how easily it knocks, precisely the opposite of the octane rating. Alcohol ignites under compression at temperatures approximately 60 percent greater than diesel fuel does. Alcohol also does not have the lubricity needed for the injectors and injector pumps in a diesel engine. As anyone who's handled diesel fuel knows, if anything, it's oily. And, the diesel injector pumps do not tolerate water.

Still, alcohol has been used proportionally in diesel-fueled engines, and research continues in this field, particularly for the fact that diesel emissions can be significantly reduced with the addition of alcohol in combustion. Back in the fuel-crisis days of the late 1970s and early 1980s, so-called fumigation systems were popular in agricultural circles. These aftermarket adaptations used carburetors mounted on the engine's intake manifold. After adjusting the injection system to cut back fuel delivery and tying in the carburetor's throttle valve to match the tractor's load, the engine essentially took between 30 and 50 percent of its fuel charge from the carburetor, depending on load. At idle, the engine consumed diesel almost exclusively.

For turbocharged diesels, an alcohol injection system was developed at around the same time by an outfit called M&W Gear Company. This used an alcohol injector mounted downstream from the turbocharger. It discerned turbocharger pressure and delivered the appropriate amount of fuel to lower diesel consumption and boost horsepower.

Some years ago, in São Paulo, Brazil, Mercedes-Benz engineers were eager to show me their ethanol-fueled diesel buses, which they'd developed as part of the state alcohol program. Using a cetane booster made from cane-based alcohol tri-ethylene glycol di-nitrate they blended a small percentage with straight ethanol and used the fuel in city buses.

To address the lubrication issue, the injection pumps were fitted with a separate filtered oil line from the oil pump and a return line to the crankcase. A small amount of oil actually gets into the fuel to help lubricate the injectors, but not enough to impact the reservoir in the crankcase.

Early in 2008, the Swedish automotive-tech company BSR Svenska AB unveiled their ethanol-powered diesel conversion, a Saab 9-3optimized for E-95 (95 percent ethanol, 5 percent gasoline) fuel. The engine's combustion chamber was modified and the fuel system and ECU software altered to suit the properties of alcohol. The 195-horsepower engine achieved fuel mileage figures of around 46 miles per gallon and reduced carbon dioxide and hydrocarbon emissions by approximately 95 percent. This automobile is not in production, but has provided a clear path for future development. The company markets E-85 conversion kits for a variety of gasoline cars.

Other experimenters have blended ethanol and plant oils to make diesel fuel with varying success. Mixing hydrated alcohol with castor oil was not an uncommon practice among fuel alchemists some 30 years ago. Blended in a 1:4 ratio with ethanol dominant, the fuel was passably acceptable though difficult to keep consistent. There is no documentation that I'm aware of that indicates its effect on engine or component longevity. More recently, blenders have combined biodiesel and anhydrous alcohol, which allows a fairly broad range of proportion. This also addresses the lubricity issue of using ethanol, and the problems associated with water in hydrated alcohol. The prudent among us will approach these experiments with eyes wide open, because of the potential damage that unsuitable fuels can wreak on costly diesel engines.

8.12   Space Heating Systems

Kerosene, or fuel oil, is still a major component of the heating industry and more common in some parts of the country than natural gas or propane. Heating fuel has the same characteristics as diesel fuel, so it burns with a distinct petroleum odor and leaves an oily residue when it's done. But fuel oil burners can easily be adjusted to burn alcohol, and with considerably less effort than it takes to modify a vehicle engine. The burners in oil furnaces are called gun burners because they're simply barrels fitted with a nozzle that injects pressurized fuel into a refractory-lined combustion crucible. An electric spark ignites the fuel spray, and the resultant heat is directed through tubes in a heat exchanger. Combusted gases go up the flue pipe and heated air is forced into the duct plenum.

The burner conversion process is very simple. Because alcohol already contains oxygen, you have to reduce the amount admitted to the burner by turning the air control down to limit draft. It may be a slotted collar or a rotating plate, but the point is to cut down the air supply to replicate the normal flame pattern. Then, because ethanol has a lower heating value in Btu's than fuel oil, you have to increase the size of the metering nozzle to get an equivalent amount of heat output.

Unthread the nozzle with a box wrench and take it out. You won't have to drill this orifice, because burner nozzles are available in several sizes, based on gallon-per-hour flow rates. Get one with a diameter about 15 percent larger, or the equivalent of a 35 percent increase in rate. Some conversions are more effective when the nozzle's spray angle is changed as well.

Though each manufacturer's wares are different, one constant is that the oil pump will probably not get the lubrication it needs from alcohol fuel alone. For this reason, it's necessary to add up to 5 percent kerosene or fuel oil to the new fuel to provide lubricity for the pump. Advocates of renewables may substitute biodiesel for the petroleum-based fuel.

R. F. (2009). *Alcohol Fuel A Guide to Making and Using Ethanol as a Renewable Fuel.* Canada, New York: New Society. Retrieve from: www.newsociety.com

Glossary

Term	Definition
Acid hydrolysis	A chemical reaction in which acid is used to convert starch or cellulose to sugar. It is often the first step in ethanol production.
ADM Hamburg AG	A division of ADM based in Germany that focuses on the production of biodiesel from rapeseed and grain.
Advanced biofuel	Also called second generation biofuels. Any biofuel produced from a sustainable feedstock that does not threaten the food supply.

AGE 85	Aviation fuel containing 85% ethanol and 15% biodiesel that can be used in piston engine aircraft. Abbreviation for Aviation Grade Ethanol 85
Alcohol	Any organic compound that contains an oxygen single-bonded to a hydrogen atom (OH group).
Algae	A group of diverse, generally autotrophic organisms that can be either single-cell or multi-cell. Various species of algae are used to produce ethanol and butanol.
Algenol	A U.S. biofuel company that produces ethanol directly from algae

	without harvesting them.
Anaerobic digestion	Breakdown of organic matter in an environment where oxygen is NOT present, which produces methane and CO_2.
Aromatics	Any hydrocarbon that contains a ring (usually six carbon atoms) in which the carbon atoms all share electrons equally, resulting in bonds with characteristics that are intermediate between single and double bonds. Bonds are considered to be unsaturated.
Australian Biofuels Research Institute (ABRI)	An organization funded by the federal government of Australia with the goal of developing biofuel

	research
B100	Biodiesel that contains no petroleum products. It is 100% renewable, organic biodiesel. Requires extensive engine modifications. VW produces B100 compatible vehicles. Prone to low-temperature gelling.
B2	A mixture of 2% biodiesel to 98% petrodiesel
B20	A mixture of 20% biodiesel to 90% petrodiesel
B5	A mixture of 5% biodiesel to 95% petrodiesel. A common blend in the UK.
Bagasse	The dry, fibrous remnants

	of sugarcane or sorghum once the sugar has been extracted. It contains mostly cellulose, hemicelluloses, and lignin. It can be burned directly to generate electricity or converted to ethanol.
Bio-butanol	Butanol produced from biomass. It can be used in unmodified gasoline engines.
Biodiesel	Diesel fuel produced from biomass. Biodiesel cannot be used in standard engines without modification as it corrodes rubber seals and gaskets. It also has a lower gelling point than petro diesel, making it unsuitable for use in colder climates.

	Biodiesel is often blended with petro diesel.
Biodiversity	The amount of variation in an ecosystem. Greater variation indicates a healthier ecosystem. Genetically modified organisms and large monocultures used in biofuel production can threaten biodiversity.
Bio-energy	Any renewable energy made from biological sources. Fossil fuels are not counted because, even though they were once biological, they are long dead and have undergone extensive modification.
Biofuel	Any fuel derived from biological carbon fixation,

	including solid fuels, bioethanol and other bio-alcohols, biodiesel, etc.
Biogas	A mixture of methane and CO_2 with water vapor created by the breakdown of organic matter in the presence of oxygen.
Bio-hydrogen	Hydrogen that is produced biologically (most often by bacteria and algae).
Bioreactor	Any device that supports a biologically active environment. In the context of biofuels, a bioreactor is most often used to grow algae.
Black liquor	Used cooking liquor from the kraft process for producing paper. It

	contains lignin, hemicelluloses, and other wood extracts. It is considered a waste product and can be gasified to produce biofuel.
Blue Marble Bio-material	A U.S. biofuel company that produces biofuel and biochemical from various feedstock.
British Thermal Unit (BTU)	A measure of heat that is equivalent to the amount of energy required to raise the temperature of 1 pound of water by 1 degree Fahrenheit.
Brown liquor	Similar to black liquor, but it is derived from the sulfite process of paper production. Alternatively

	called red liquor, thick liquor, and sulfite liquor.
Butamax Advanced Biofuels	A joint venture between BP and DuPont to produce bio-butanol from sugar and starch feedstock.
Butanol	A six-carbon alcohol that is more similar to gasoline than ethanol. It can be used in gasoline engines without the need for modification and produces roughly that same amount of energyper mass as gasoline.
Camelina	A genus of flowering plant related to flax. Some species produce seeds with large quantities of oil that can be converted to various biofuels (e.g.

	biodiesel). Camelina grows well in moderate climates.
Carbon dioxide	A molecule made of one carbon atom double bonded to two oxygen atoms (one of each side of the carbon). It is a colorless, odorless gas at standard temperature and pressure and is widely implicated as one of the major causal agents in greenhouse warming.
Carbon monoxide	A molecule made of one carbon atom bonded to a single oxygen atom via a triple bond. It is a product of inefficient combustion of hydrocarbon compounds. It is a pollutant and is toxic to

	humans at concentrations above 50 ppm during long term exposure or 667 ppm during short term exposure.
Carbon Negative	A product or process that, over its entire lifetime, causes a net decrease in atmospheric carbon levels.
Carbon Neutral	A product or process that, over its entire lifetime, causes no net increase or decrease in atmospheric carbon levels.
Carbon Positive	A product or process that, over its entire lifetime, causes a net increase in atmospheric carbon levels.
Carbon sink	Any reservoir that can accumulate or store

	carbon-containing compounds for a prolonged or indefinite period. This is particularly relevant to carbon dioxide, which could be stored to reduce its environmental impact.
Cassava	A woody, tropical shrub well known for being the source of tapioca extract. It is used in southeast Asai for the production of biodiesel and ethanol.
Cellulose	An organic polysaccharide of the general formula $(C6H10O5)n$ that is the structural component of the cell wall of most green plants. It is used in several biofuel production processes. Biofuel

	produced primarily from cellulose is sometimes called cellulosic biofuel.
Cellulosic ethanol	Ethanol produced from the inedible, cellulose-rich parts of plants. Studies are conflicting on its benefits, but he U.S. Department of Energy has suggested it can reduce greenhouse gas emissions by as much as eighty-five percent over reformulated gasoline.
Cetane Rating	A number used to rate the quality of diesel fuel or any fuel combusted via compression ignition. Petro diesel ranges from cetane 40-55 in most cases while biodiesel ranges from 46-52 if plant-based

	and 56-60 if animal based
Chaff	The protective casings of seeds of many grains, cereals, and straw. Some livestock can consume it, but it is generally considered to be a waste product.
Charcoal	A residue consisting of amorphous carbon that is obtained by pyrolysis or burning of wood or other organic compounds. It is used in the production of syngas.
Chemrec	A company based in Stockholm, Sweden that specializes in the conversion of black and brown liquors to syngas

Chinese National Petroleum Corporation	The largest stated-owned oil and gas company in China. CNPC has recently branched into ethanol production to meet the expanding demands of the Chinese economy for fuel.
Cloud Point	The temperature at which solids dissolved in al liquid are no longer completely soluble and begin to precipitate. Biodiesel often reaches a cloud point at higher temperatures than petro-diesel, making it less suitable for cold environments. This is one measure of the quality of diesel fuel or aviation fuel.
Cogeneration	The production of electric energy along with a

	second for of energy (often heat).
Combined Heat and Power	See Cogeneration
Combustion	Commonly referred to as burning. This is the process by which a fuel and an oxidant react to product heat (energy) and other compounds (CO_2 and H_2O in ideal hydrocarbon combustion)
Combustion Gases	The gases released during a combustion process. Similar to emissions.
Compression-ignition engine	An internal combustion engine in which the fuel is ignited by heat generated from compressing the gas to high pressures rather

	than from a spark. Most diesel engines work this way.
Corn	Also called maize, corn is a domesticated grain grown widely throughout the world. Corn is the largest source of ethanol feedstock in the world. Forty percent of the 332 million metric tons of corn grown in the U.S. is used to make ethanol.
Cunningham Ella japonica	A type of tropical fungus recently found to produce hydrocarbons from organic matter. The hydrocarbons indistinguishable from those found in fossil fuels.
DDG	See Dried Distillers Grains

DDGS	See Dried Distillers Grains with Soluble
Denatured ethanol	Ethanol that has additives to make it undrinkable. It is commonly used as a fuel and in chemical applications.
Diesel fuel	Any liquid fuel used in a diesel engine.
Diester	The French term for biodiesel. It is a contraction of the words diesel and ester.
Diester Industrie	A subsidiary of the French petroleum company Diester Industrie International that specializes in the production of biodiesel

Dimethyl furan	A heterocyclic compound that can be used as a biofuel. It is easier to produce than butanol and has an energy content 40% greater than ethanol.
Direct Land Use Changes	Refers to changes in the way humans impact a given area of the earth and how that change affects greenhouse gas emissions. Attention is given to water diversion, biodiversity, and limited energy investment costs.
Distiller's Wet Grains (DWG)	Grains containing more than 12% water and often up to 70% water.
Dried Distillers Grains	Unfermented grain residues that have been dried so as to extend the

	shelf life.
Dried Distillers Grains with Solubles	Unfermented grain residues that have been dried to 10-12% moisture so as to extend shelf life. Commonly used in ethanol production
Dry milling	Mechanical grinding of a feedstock that does not use water.
Dryer	Apparatus that removes water from distiller's wet grains to produce distiller's dried grains.
DuPont Danisco	A joint venture between DuPont and Danisco foods to produce cellulosic ethanol using sugarcane, corn, and switch grass.

E10	A blend of 10% ethanol and 90% gasoline. The standard fuel in the United States
E100	100% fuel ethanol. Used mainly in Brazil.
E2	A blend of 2% ethanol and 98% gasoline.
E85	A blend of 85% ethanol and 15% gasoline, which is the common blend implied by the term flex fuel vehicle (FFV). Generally seen only in the Brazil and the Midwestern United States.
E90	A blend of 90% ethanol and 10% gasoline.
Emissions	The gaseous or particulate

	components expelled during a combustion reaction. The term commonly refers to the mix of gases and particulate that exit the exhaust of an internal combustion engine.
Energy Balance	In regard to biofuel, this term refers to the amount of energy required to produce the fuel versus the amount of energy derived from the fuel.
Energy Content	Also referred to as heating value, energy content is a measure of the number of British Thermal Units obtained by burning a set volume of fuel. Because it relies on volume, energy content can change with

	temperature and pressure.
Energy Crop	A low-cost, low maintenance plant gown exclusively for use as fuel.
Energy Density	Generally, the amount of energy stored in a given region of space per unit volume. Specifically, the amount of energy obtained from a specified mass of biofuel. Useful for comparing various types of biofuels in a standardized manner.
Energy Efficiency Ratio	A comparison of the energy stored in a fuel and the energy needed to produce, transport, and distribute the fuel.
Enzyme	A biological molecule that

	performs chemical inter conversions. I.E., a molecule from a living organism that converts one chemical into another.
Ethanol	An alcohol composed of two carbons. The formula is C2H4O
Ether	A class of organic compounds that contains an ether group defined as R-O-R' where R is any hydrocarbon chain and R' is another hydrocarbon chain. Ethers are under consideration as biofuels and are used as additives in current fossil fuel blends
Ethyl Alcohol	See ethanol

Ethyl Tertiary Butyl Ether (ETBE)	Ether commonly used to oxygenate gasoline. Ether provides for cleaner combustion and decreased emissions.
FAME biodiesel	Fatty Acid Methyl Ester biodiesel
Feedstock	The raw material from which a biofuel is produced. Feedstock is generally a plant itself, but in the case of algae, the feedstock is any source of carbon (often carbon dioxide)
Fermentation	A normal metabolic process in which oxygen is NOT the final electron acceptor. Fermentation is the process through which alcohols are made.

FFV	See Flexible Fuel Vehicle.
First generation biofuel	Any biofuel derived from sugar, starch, or vegetable oil. In general, these fuels are considered a threat to food supply chains.
Fischer-Tropsch Biodiesel	Biodiesel made using the Fischer-Tropsch process
Fischer-Tropsch Process	A series of chemical reactions that convert carbon monoxide and hydrogen into liquid hydrocarbons. It is used in the production of synthetic lubricants and low-sulfur diesel fuel.
Flashpoint	The lowest temperature at which a flammable liquid produces enough vapor to ignite. For most

	flammable liquids like gasoline, it is the vapor and not the liquid itself that is combustible.
Flexible Fuel Vehicle	A vehicle with an internal combustion engine that can run on more than one fuel. Usually the vehicle is designed to run on pure gasoline or a defined blend of gasoline/ethanol or gas online/methanol. In some cases the vehicle can run on pure ethanol.
Fossil fuel	A fuel formed by the anaerobic decomposition, over thousands or millions of years, of dead biomass. Petroleum, coal, and natural gas are major categories of fossil fuels.

Fungi	One of the kingdoms of living organisms. Fungi are eukaryotic organisms that are similar to plants in many ways, but which are not able to perform photosynthesis. Yeast, mold, and mushrooms are all fungi.
Gas to liquid	A refinery process that takes gaseous hydrocarbons like natural gas and converts them into longer hydrocarbons that are liquids. Methane and syngas are commonly converted to liquid fuels.
Gasification	The conversion of carbonaceous solids into carbon monoxide, hydrogen, and carbon dioxide gases. This usually

	takes place at high temperatures and in the presence of steam.
Gasohol	Blends that are between E5 and E25 are commonly referred to as gasohol in the United States. The E10 blend is most commonly called gasohol throughout the rest of the world.
Gel point	The temperature at which an infinite polymer network is formed. Sometimes this term is used interchangeably with cloud point, but the two do not mean the same thing. At cloud point, solute precipitates (comes out of) from the solution whereas at gel point, the components of the

	substance form polymer chains long enough to make the substance solid rather than liquid. Gel point is also defined as the point that a liquid fuel takes on the consistency of petroleum jelly.
Genetically modified organism (GMO)	A living thing that has had its genetic material (either DNA or RNA) manipulated by humans using recombinant techniques. The *direct* modification of DNA through an artificial process.
Gevo	A U.S. company that creates "renewable chemicals" and isobutanol using genetically modified

	microorganisms.
Gliocladium roseum	A fungus from Patagonia that produces diesel fuel from cellulose.
Glucose	The sugar most commonly produced by living organisms.
Glycerin	See Glycerol
Glycerol	A simple compound mad of three carbons and three OH (hydroxyl) groups. It is a component of fatty acids and, as a result, a byproduct of biodiesel production via trans-esterification.
Grain Alcohol	See Ethanol
Green diesel	See Renewable Diesel

Greenhouse Effect	The warming of the Earth as a result of heat that is trapped by certain gases.
Greenhouse Gas	Any gas that traps heat in the Earth's atmosphere and thus leads to increased temperatures. Water vapor, carbon dioxide, methane, ozone, chlorofluorocarbons, and nitrous oxide are all greenhouse gases.
Gushan Environemental Energy	A Chinese biodiesel and glycerol producer that uses vegetable oil and used cooking oil for feedstock.
Hydrocarbon	A compound that consists solely of hydrogen and carbon atoms. Most fossil fuels are hydrocarbons.

Hydrogen sulfide	A poisonous, foul-smelling (rotten eggs) gas mde up of two hydrogen atoms and one sulfur atom. It is often found along with petroleumdeposits and is considered a contaminant in fossil fuels.
Indirect Land Use Changes	Changes to land as a result of growing biofuel feedstock that are not a direct result of human intervention. This can include loss of biodiversity, subsequent changes to the ecosystem that have broader impact, and more. Essentially, this refers to the unintended consequences of releasing more CO_2 by using land for biofuel feedstock growth. Abbreviated

	ILUC.
International Energy Agency (IEA) Bioenergy	A subdivision of the IEA working to achieve global integration of substantial bioenergy use.
Jatro BioJet Fuel	A German company focusing on the production of Jet Fuel from Jatropha. The company produces certified Jet A-1 fuel, which has been using in several test flights and is set to be purchased by several major airlines in the coming years.
Jatropha	A genus of flowering plants that grown in tropic regions, on marginal land. Oil from Jatropha seeds can be used to produce

	biodiesel.
Jilin Fuel Ethanol	A Chinese bioethanol company that operates the largest bioethanol plant in the world.
Joule	A measure of energy or work defined as the energy expended by applying a force of one newton over a distance of one meter.
Joule Unlimited	A U.S. biofuel company that focuses on the production of ethanol and biodiesel using genetically modified algae.
Knock	A pinging sound that occurs in internal combustion engines when fuel is burned at the incorrect time. This often

	occurs because the fuel either burns too quickly, too slowly, or too early in the cycle. Fuels with a low octane rating are more prone to detonate early or burn too long.
Landfill Gas	Biogas (methane, CO_2, and water vapor) produced from the breakdown of organic material in landfills.
Lifecycle analysis	Determining the environmental impact of all stages of an activity starting with raw material extraction through production, use, and disposal.
Lignin	A complex chemical found mostly in woody plants

	and some algae
Lignocellulose	A group of substances that make up the plant walls of woody plants and which consist of a mixture of cellulose and lignin.
LS9 Inc.	A U.S. company that produce biofuels from genetically modified organisms, primarily algae
Lubricity	A measure of how well a lubricant reduces friction. Also referred to as a lubricant's "anti-wear property."
M85	A mixture of 85% methanol and 15% gasoline. It is relatively uncommon given the low energy density and high

	toxicity of methanol.
Mash	A mixture of grain and water that is used as the base for fermenting ethanol.
Mega joule	1,000,000 joules. A common measure of energy in fuels.
Methanol	Also called wood alcohol. Methanol is composed of a single carbon, one oxygen, and four hydrogen molecules. It is highly toxic and has a relatively low energy density.
Methyl Ester	The primary component of biodiesel, created by adding methanol to a fatty acid to produce three methyl esters and a

	glycerol molecule.
Methyl Tertiary Butyl Ether (MTBE)	A common gasoline additive used to raise octane and thus reduce knock. It is used to oxygenate gasoline.
Multi-feedstock	A technology that is able to produce fuel or energy from more than one type of feedstock.
National Fuel Alcohol Program	Also called the Programma Nactional do Alcool in Brazil, this was the national policy of the country of Brazil that set standards for the inclusion of ethanol in fuel. As a result of this program, Brazil became the global leader in ethanol-based fuels.

National Renewable Energy Laboratory (NREL)	The scientific arm of the U.S. government tasked with research and development of alternative energy technology.
Neat Fuel	Any fuel that is pure or unmixed. For example, neat ethanol is 100% ethanol.
Nitrogen	Nitrogen is the name of both an element and a molecule. Molecular nitrogen is a colorless, odorless gas made up of two nitrogen atoms covalently bonded to one another. Under the high temperatures and pressures of combustion, nitrogen can combine with other compounds to form nitrogen oxides, which are

	considered pollutants.
Nitrogen oxides	A group of compounds containing oxygen and nitrogen in varying ratios. NO and NO2, generically referred to as NOx, are produced during high temperature combustion and contribute to the production of acid rain and ozone.
Novozymes	A biotech company headquartered in Denmark that produces enzymes and microorganisms for various industries, including biofuels.
Octane	A collection of hydrocarbons of the formula C8H12 that are

	components of gasoline. Pure isooctane, one of the structural isomers of octane, is used as the reference for rating the anti-knock qualities of gasoline.
Octane Rating	A way of rating the performance of motor and aviation fuels. Higher octane numbers indicate the ability of a fuel to withstand greater compression without detonation (burning). Higher compression generally means higher engine performance.
OPEC	Oil Producing & Exporting Countries
Organization for	An international

Economic Cooperation and Development (OECD)	association of 34 countries that work to improve the economic situation of all people on the planet. Part of their focus includes holding down the costs of energy and so they have a lot of input regarding biofuel policy.
Palm oil	An edible vegetable oil derived from oil palm trees. Palm oil is used to produce biodiesel.
Particulate	A small quantity of solid or liquid that is dispersed within a gaseous (or liquid) emission. Dust, smoke, soot, aerosols, and sprays are all particulates.
Perennial	A plant that lives more than two years.

Petrobras	A semi-public petroleum corporation headquartered in Brazil. Petrobras is partially controlled by the state and so makes decisions regarding the Brazilian Ethanol Program.
Photo-bioreactor	A device that supports a biologically active environment that incorporates a light source to provide energy. Often used in the growth of algae.
Photosynthesis	The process by which living organisms like plants and algae convert light energy into chemical energy.
Programma	See National Fuel Alcohol

Nactional do Álcool	Program
Proof	In reference to ethanol, a proof is equivalent to 0.5% by volume. For example, a 30 proof alcoholic beverage would be 15% alcohol by volume.
Rapeseed	A flowering member of the mustard/cabbage family that grows in temperate climates.
Reformulated gasoline (RFG)	Gasoline blended to burn more cleanly and to reduce smog and other toxins from entering the air. Reformulated gasoline contains MTBE, ETBE, or an alcohol to add oxygen for improved combustion.
Renewable Diesel	Any diesel fuel that is

	produced from a renewable source. Renewable diesel does not have to be environmentally friendly or reduce GHG emissions.
Renewable Fuels Standard (RFS)	The U.S. policy governing the minimum amount of renewable fuel in transportation fuel.
Sapphire Energy	A U.S. biofuel company that produces crude oil from algae. The oil can be refined to generate gasoline, jet fuel, and diesel fuel.
Second generation biofuel	Also called advanced biofuels. Any biofuel produced from a sustainable feedstock that does not threaten the food

	supply.
SG Biofuels	A U.S. biofuel company that focuses on molecular breeding and biotechnology to produce hybrid varieties of Jatropha.
Solazyme	A U.S. biofuel company that produces algal fuel for use in ground and air application. The company also produces personal care and nutritional products derived from algae.
Solid Biofuel	Any solid biomass including wood, sawdust, grass trimmings, charcoal, agricultural waste, and dried manure.

Solix Biofuels	A U.S. company that focuses on the production of modular growth solutions for algae used in biofuel production.
Soybean	A type of legume native to Asia that is used primarily as a food, but which also has limited application in biofuel production.
Starch	A polymer of glucose molecules that is made by plants. Starch must be converted to sugar (individual glucose molecules) before it can be fermented to produce alcohol.
Stover	The dried stalks and leaves of a crop after the edible parts have been

	harvested.
Sugarcane	Any of 37 species of perennial grass rich in sugar and native to warm tropical regions. Sugarcane is the primary source of ethanol in Brazil, yielding roughly 800 gallons of ethanol per acre of planting.
Sulfur	Chemical element number 16. This abundant element is a common contaminant in fossil fuels which, when burned, produce sulfur dioxides that eventually become acid rain.
Switch grass	A perennial grass native to North America that is considered as a potential fuel for the production

	electricity via direct combustion.
Syngas	A mixture of carbon monoxide and hydrogen that can be used to produce synthetic petroleum via the Fischer-Tropsch process.
Thermal Conversion	The transformation of complex organic material into light crude oil using pressure and heat.
Trans-esterification	The process of exchanging the organic group of an ester with the organic group of an alcohol. The process is used to produce biodiesel from triglyceride.
Triglyceride	Organic compounds

	composed of three fatty acid chains connected to a glycerol molecule.
Vegetable Oil	A triglyceride extracted from a plant.
Viscosity	A measure of the resistance of a fluid to deformation by shear stress or tensile stress. Viscosity is an important component of lubricating oils.
Volatility	A measure how easily a liquid is converted to a gas. The lower the temperature at which a liquid evaporates, the more volatile it is considered to be.
Waste Vegetable Oil	Vegetable oil that is no

(WVO)	longer fit for use in food.
Wet milling	A process in which feedstock is steeped in water to soften it before it is ground.
Wood Alcohol	See Methanol.
Zero Carbon Contribution	See Carbon Neutral

Biofuels - Glossary of Terms. (n.d.). Retrieved from http://biofuel.org.uk/glossary.html

INDEX

B

C

REFERENCE

- R. F. (2009). *Alcohol Fuel A Guide to Making and Using Ethanol as a Renewable Fuel.* Canada, New York: New Society. Retrieve from: www.newsociety.com

- FUEL FROM FARMS A Guide to Small-Scale Ethanol Produ.ction Second Edition

- Nathan, R. A. Fuels from Sugar Crops. DOECritical Review Series. 1978. Available from NTIS, #TID-22781.

- Stauffer, M. D.; Chubey, B. B.; Dorrell, D. G.

Jerusalem Artichoke. A publication of Agriculture Canada, Research Station, P. 0. Box 3001, Morden, Manitoba, ROC 1 JO, Canada. 1975.e

www.ingramcontent.com/pod-product-compliance
Lightning Source LLC
Chambersburg PA
CBHW072131170526
45158CB00004BA/1329